More Calculus

Problem Solving using the *fx-991ES* Calculator

Dr Allen Brown

Cambridge
Paperbacks

Cambridge Paperbacks

www.CambridgePaperbacks.com

First published by Cambridge Paperbacks 2019

ISBN: 978-1-9161856-3-0

Disclaimer

Although the author and publisher have made every effort to ensure that the information in this book was correct during preparation and printing, the author and publisher hereby disclaim any liability to any party for any errors or omissions.

Read this First

Differentiation is an important part of any calculus course and the only way of becoming proficient is through practice. In this book there are many examples of using the *chain rule* and the *quotient rule* for performing differentiation. Whenever performing a differentiation, you are not quite sure whether the result you have is correct. The main feature which makes this book particularly attractive is the use of the *fx-991ES* calculator for *confirming* your result. In every example the keystrokes for your calculator are given, all you have to do is repeat the same keystrokes and you will get the correct answers. The more times you do this, the more confident you will become at using your calculator as a real resource in helping you to understand calculus.

Also, learning integration is not an easy task and requires a lot of personal application. However, as with differentiation, the more you practice the better you become at solving integration problems. To make the learning easier, this book contains lots of worked examples. Having performed an integration, you will want to know if your answer is correct. This book shows you how to use your *fx-991ES* calculator to check your answers. Not only do you learn about integration, but also how to use efficiently your scientific calculator for solving really difficult problems. This book therefore teaches you how to use your *fx-991ES* calculator to confirm you have performed the integration correctly.

You are expected to work through each example yourself and key-in all the keystrokes to gain confidence in problem solving. Once you have gained a good working skill you can apply your new skill set to your class-room calculus problems.

Dr Allen Brown
Cambridgeshire

Contents

1 Differentiation Examples

There are twelve fully worked examples in this first half of this book. Work through each example and use your *fx-991ES* calculator to confirm your results.

Example 1:

Differentiate the expression

$$y = \frac{(1+x)^2}{1+2x}$$

You will know whenever you have an expression like this the quotient rule is used,

$$y = \frac{u}{v}$$

$$\frac{dy}{dx} = \frac{1}{v^2}\left\{v\frac{d}{dx}u - u\frac{d}{dx}v\right\}$$

$$\frac{dy}{dx} = \frac{1}{(1+2x)^2}\left\{(1+2x)\frac{d}{dx}(1+x)^2 - (1+x)^2\frac{d}{dx}(1+2x)\right\}$$

Perform the differentiation,

$$\frac{1}{(1+2x)^2}\{(1+2x)2(1+x) - 2(1+x)^2\}$$

$$\frac{2}{(1+2x)^2}\{(1+2x)(1+x) - (1+x)^2\}$$

$$\frac{2}{(1+2x)^2}\{1+3x+2x^2 - (1+x)^2\}$$

$$= \frac{2}{(1+2x)^2}\{1 + 3x + 2x^2 - 1 - 2x - x^2\}$$

Collect terms

$$\frac{2}{(1+2x)^2}\{x + x^2\}$$

Leaving

$$\frac{dy}{dx} = \frac{2x(1+x)}{(1+2x)^2}$$

Evaluate this expression with three x test values, -0.13, 2.2, 4.1

The keystrokes for your *fx-991ES* calculator are,

[▭/▭] [2] [ALPHA] (X) [(] [1] [+] [RCL] (X) [)] [▼] [(] [1] [+] [2] [RCL] (X) [)] [x^2]

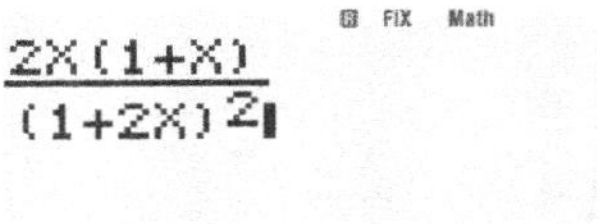

The first value is -0.13

[CALC] [(−)] [0] [•] [1] [3] [=] [S⇔D]

The second value is 2.2

[CALC] [2] [•] [2] [=] [S⇔D]

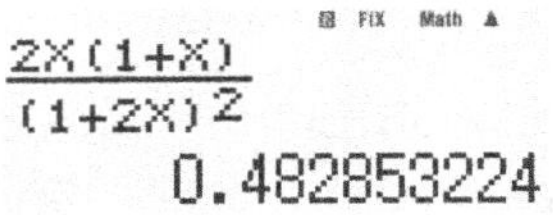

The third value is 4.1

[CALC] [4] [•] [1] [=] [S⇔D]

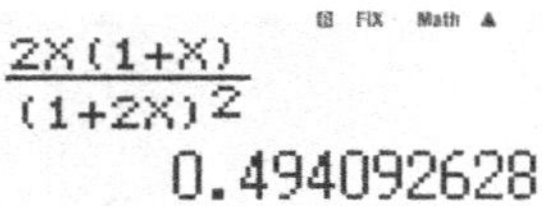

The original expression is,

$$y = \frac{(1+x)^2}{1+2x}$$

Use the differentiation feature on your calculator, the keystrokes are,

[SHIFT] ($\frac{d}{dx}$■) [▭/▭] [(] [1] [+] [ALPHA] (X) [)] [x^2] [▼] [1] [+] [2] [RCL] (X) [▶] [▶]

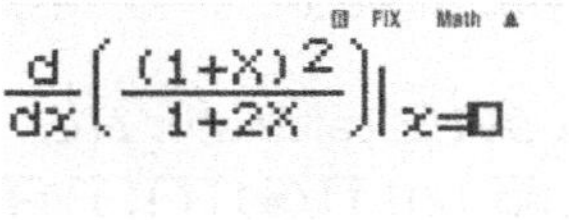

The first test value -0.13

[(−)] [0] [•] [1] [3] [=]

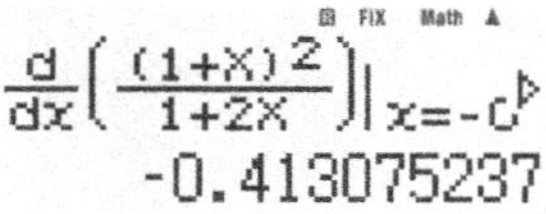

Which matches the calculation on the previous page. Enter the following keystrokes for the second test value, [◀] [DEL] [DEL] [DEL] [DEL] [DEL] [DEL]

The second test value 2.2

[2] [•] [2] [=]

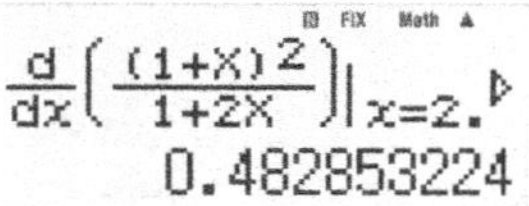

Which matches the calculation on the previous page. Enter the following keystrokes for the third test value, [◀] [DEL] [DEL] [DEL] [DEL]

The third value 4.1

[4] [·] [1] [=]

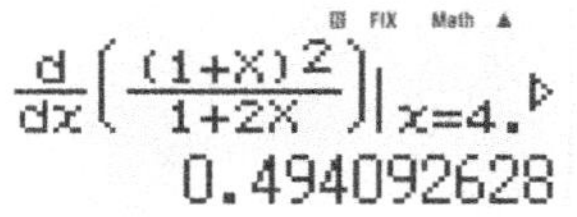

Which corresponds to the result on page 4. A graph of the function and its derivative is shown on the next page.

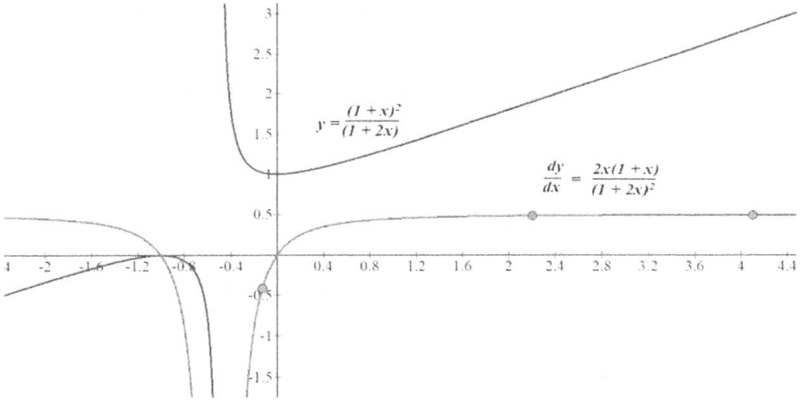

The asymptote is present at x = -0.5 and the three test values are shown in red on the derivative plot.

Example 2:

Differentiate the expression

$$y = \sin(3x)\cos^3(x)$$

Use the product rule,

$$\frac{dy}{dx} = \sin(3x)\,\frac{d}{dx}\cos^3(x) + \cos^3(x)\frac{d}{dx}\sin(3x)$$

Let $u = \cos(x)$, then $\frac{du}{dx} = -\sin(x)$

and

$$\frac{d}{dx}u^3 = 3u^2\frac{du}{dx}$$

$$\frac{d}{dx}\sin(3x) = 3\cos(3x)$$

$$\frac{dy}{dx} = \sin(3x)\frac{d}{dx}\cos^3(x) + \cos^3(x)\frac{d}{dx}\sin(3x)$$
$$= \sin(3x)\, 3u^2\frac{du}{dx} + 3\cos^3(x)\sin(3x)$$

$$= \sin(3x)\, 3\cos^2(x)\, [-\sin(x)] + 3\cos^3(x)\cos(3x)$$
$$= 3\cos^2(x)[\cos(x)\cos(3x) - \sin(x)\sin(3x)]$$

Use the identity,

$$\cos(A + B) = \cos(A)\cos(B) - \sin(A)\sin(B)$$

Let $A = 3x$ and $B = x$

$$\cos(3x + x) = \cos(3x)\cos(x) - \sin(3x)\sin(x)$$

Leaving

$$\cos(4x) = \cos(3x)\cos(x) - \sin(3x)\sin(x)$$

Therefore

$$\frac{dy}{dx} = 3\cos^2(x)[\cos(x)\cos(3x) - \sin(x)\sin(3x)]$$

or

$$\frac{dy}{dx} = 3\cos^2(x)\cos(4x)$$

Test for three radian values of x,

$x = -2.1, \qquad x = 1.7, \qquad x = 3.6$

The keystrokes for your calculator,

[3] [cos] [ALPHA] (X) [)] [x^2] [cos] [4] [RCL] (X) [)]

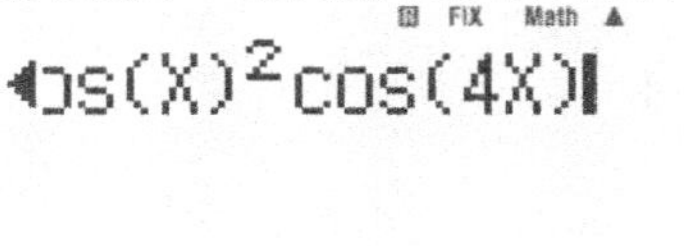

First test value, $x = -2.1$

[CALC] [(−)] [2] [•] [1] [=]

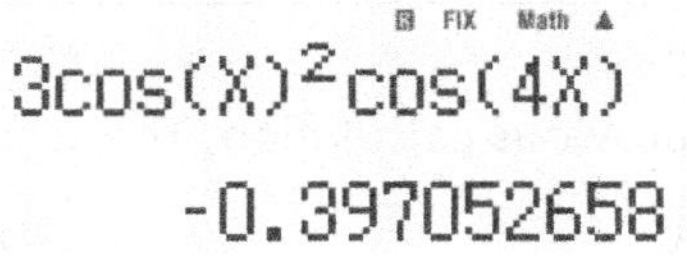

Second test value $x = 1.7$

[CALC] [1] [•] [7] [=]

Third test value $x = 3.6$

[CALC] [3] [•] [6] [=]

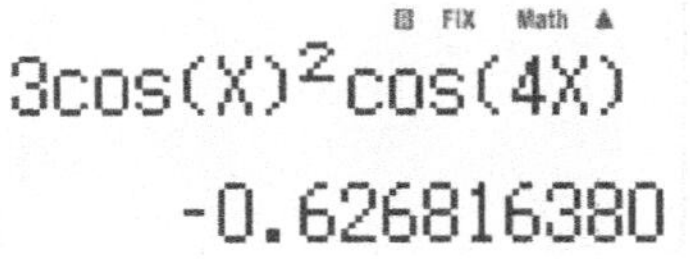

Use the differentiation feature to verify the results. The original expression,

$$y = \sin(3x)\cos^3(x)$$

The keystrokes are,

[AC] [SHIFT] ($\frac{d}{dx}$▪) [sin] [3] [ALPHA] (X) [)] [cos] [RCL] (X) [)] [SHIFT] (x^3) [▶]

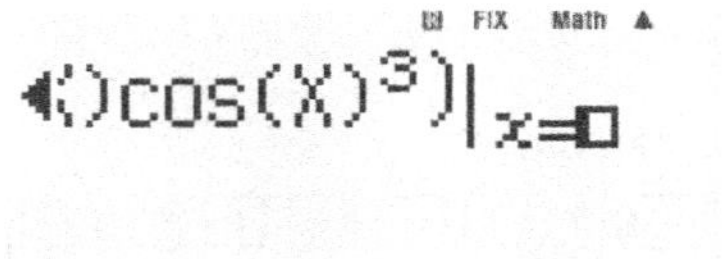

First test value $x = -2.1$

[(−)] [2] [•] [1] [=]

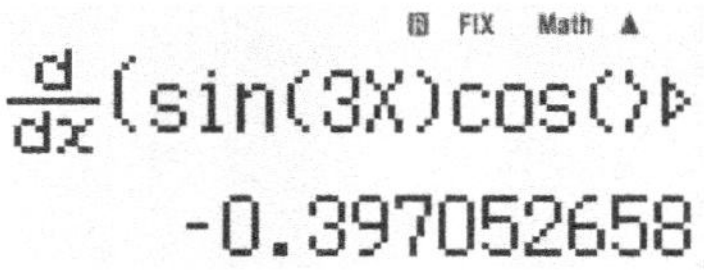

Which corresponds to previous calculation. The second test value $x = 1.7$,

[◀] [DEL] [DEL] [DEL] [DEL] [DEL]
[1] [•] [7] [=]

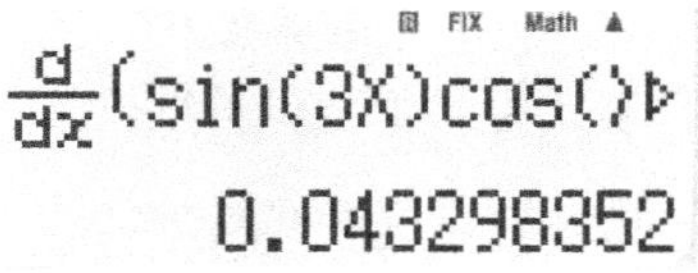

Which again corresponds to the value on page 8 and the third test value $x = 3.6$

[◀] [DEL] [DEL] [DEL] [DEL]
[3] [•] [6] [=]

which confirms the differentiation is correct. A plot of the expression and its derivative can be seen on the next page.

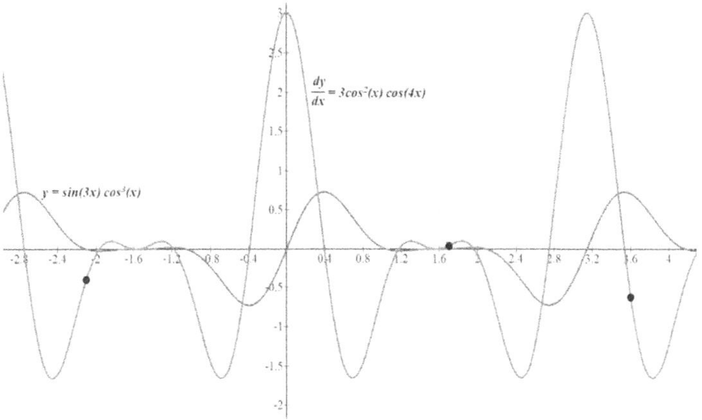

Example 3:

Differentiate the expression

$$y = \frac{x}{\sqrt{\sin(x)}}$$

Use the quotient rule,

$$\frac{dy}{dx} = \frac{1}{\sin(x)}\left\{\sqrt{\sin(x)}\frac{d}{dx}\,x - x\frac{d}{dx}\sqrt{\sin(x)}\right\}$$

$$= \frac{1}{\sin(x)}\left\{\sqrt{\sin(x)} - x\frac{d}{dx}\sqrt{\sin(x)}\right\}$$

Let $u = \sin(x)$ then $\frac{du}{dx} = \cos(x)$

$$\frac{dy}{dx} = \frac{1}{\sin(x)}\left\{\sqrt{\sin(x)} - x\frac{d}{dx}u^{\frac{1}{2}}\right\}$$

$$= \frac{1}{\sin(x)}\left\{\sqrt{\sin(x)} - \frac{x}{2}u^{-\frac{1}{2}}\frac{du}{dx}\right\}$$

$$= \frac{1}{\sin(x)}\left\{\sqrt{\sin(x)} - \frac{x}{2}\frac{1}{u^{\frac{1}{2}}}\frac{du}{dx}\right\}$$

Since $u = \sin(x)\ and\ \frac{du}{dx} = \cos(x)$

$$\frac{dy}{dx} = \frac{1}{\sin(x)}\left\{\sqrt{\sin(x)} - \frac{x}{2}\frac{1}{\sqrt{\sin(x)}}\cos(x)\right\}$$

$$= \frac{\sqrt{\sin(x)}}{\sin(x)}\left\{1 - \frac{x\cos(x)}{2\sqrt{\sin(x)} \times \sqrt{\sin(x)}}\right\}$$

$$= \frac{1}{\sqrt{\sin(x)}}\left\{1 - \frac{x}{2}\frac{\cos(x)}{\sin(x)}\right\}$$

$$\frac{dy}{dx} = \frac{1}{\sqrt{\sin(x)}}\left\{1 - \frac{x}{2}\frac{1}{\tan(x)}\right\}$$

Calculate the gradient at three test values of x,

$$x = -5.6, \qquad x = 3.1, \qquad x = 5.7$$

The keystrokes for your *fx-991ES* calculator are,

[1] [▭/▭] [√■] [sin] [ALPHA] (X) [)] ▶ ▶ [(] [1] [−] [RCL] (X) [▭/▭] [2] [tan] [RCL] (X) [)] ▶ [)]

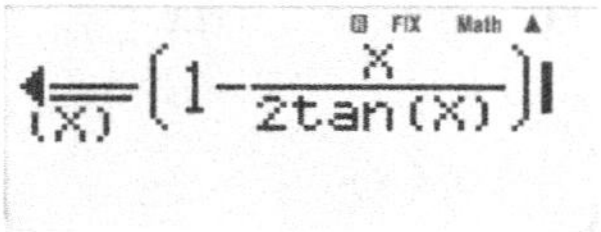

First test point $x = -5.6$

[CALC] [(−)] [5] [•] [6] [=]

Second test point $x = 1.1$

CALC 1 • 1 =

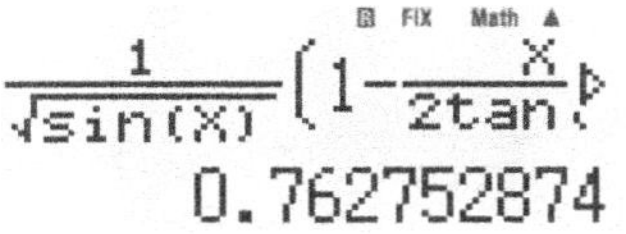

Third test point $x = 2.3$

CALC 2 • 3 =

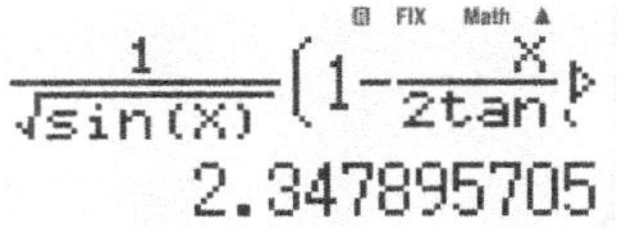

The original equation is,

$$y = \frac{x}{\sqrt{\sin(x)}}$$

Use the differentiation feature on calculator to confirm the differentiation is correct, the keystrokes are,

AC SHIFT ($\frac{d}{dx}$■) ALPHA (X) ▭ √■ sin RCL (X))

▶ ▶ ▶

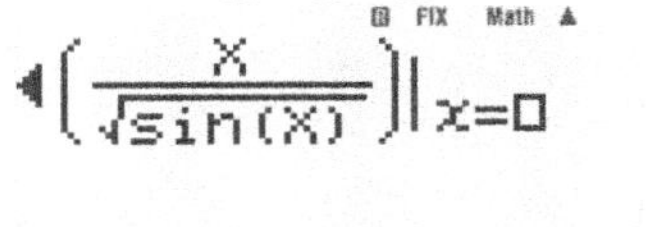

First test point $x = -5.6$

(−) 5 • 6 =

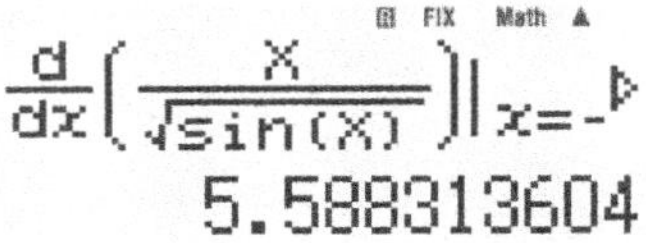

Which corresponds to the previous calculation. The second test point $x = 1.1$

◀ DEL DEL DEL DEL DEL
1 · 1 =

Which again corresponds to the previous calculation and finally the third test point $x = 2.3$

◀ DEL DEL DEL DEL
2 · 3 =

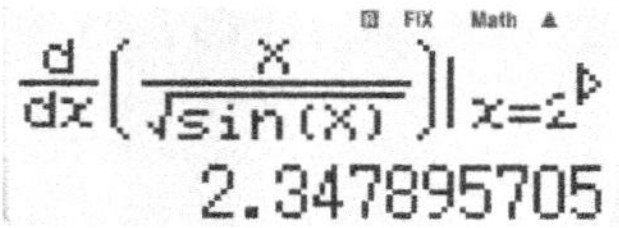

These results confirm the correctness of the differentiation. Here is a graph of the expression and its derivative,

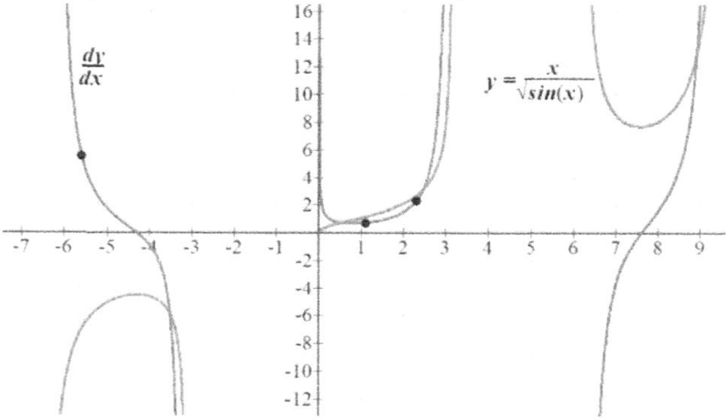

Example 4:

Differentiate the expression

$$y = \frac{1 + \sin^2(x)}{1 - \sin^2(x)}$$

Let $u = \sin(x)$ $\quad \frac{du}{dx} = \cos(x)$

$$y = \frac{1 + u^2}{1 - u^2}$$

$$\frac{dy}{dx} = \frac{1}{(1 - u^2)^2}\left\{(1 - u^2)\frac{d}{dx}(1 + u^2) - (1 + u^2)\frac{d}{dx}(1 - u^2)\right\}$$

$$= \frac{1}{(1 - u^2)^2}\left\{(1 - u^2)2u\frac{du}{dx} + (1 + u^2)2u\frac{du}{dx}\right\}$$

$$= \frac{2u}{(1 - u^2)^2}\frac{du}{dx}\{(1 - u^2) + (1 + u^2)\}$$

$$= \frac{4u}{(1 - u^2)^2}\frac{du}{dx}$$

$u = \sin(2x)$ and $\frac{du}{dx} = 2\cos(2x)$

$$\frac{dy}{dx} = \frac{8\sin(2x)\cos(2x)}{(1 - \sin^2(2x))^2}$$

$$1 - \sin^2(2x) = \cos^2(2x)$$

$$= \frac{8\sin(2x)\cos(2x)}{\cos^4(2x)} = \frac{8\sin(2x)}{\cos(2x)}\frac{1}{\cos^2(2x)}$$

$$\frac{dy}{dx} = \frac{8\tan(2x)}{\cos^2(2x)}$$

Calculate this derivative for three test values,

$$x = -3.7, x = 1.2, x = 5.8$$

The keystrokes for your *fx*-991ES calculator are,

[fraction] [8] [tan] [2] [ALPHA] (X) [)] [▼] [cos] [2] [RCL] (X) [)] [x^2]

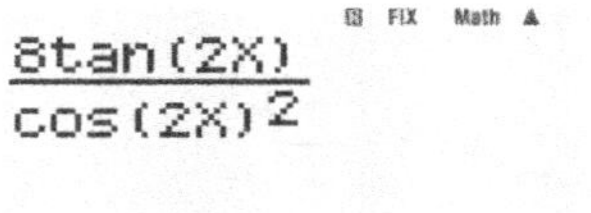

First value $x = -3.7$

[CALC] [(−)] [3] [•] [7] [=]

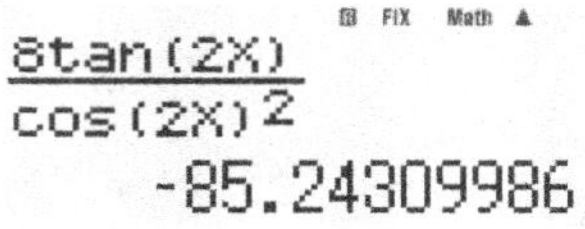

Second value $x = 1.2$

[CALC] [1] [•] [2] [=]

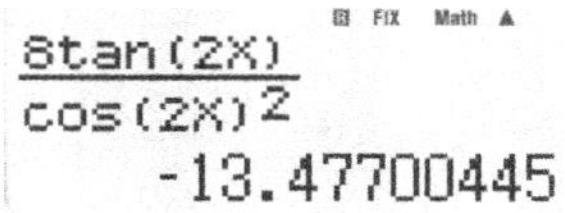

Third value $x = 5.8$

[CALC] [5] [•] [8] [=]

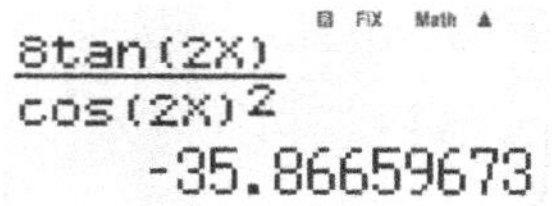

Verify these results by using the differentiation feature on your calculator, the original equation is,

$$y = \frac{1 + \sin^2(2x)}{1 - \sin^2(2x)}$$

The keystrokes are,

[AC] [SHIFT] ($\frac{d}{dx}$▪) [▭/▭] [1] [+] [sin] [2] [ALPHA] (X) [)] [x^2] [▼] [1] [−] [sin] [2] [RCL] (X) [)] [x^2] [▶] [▶]

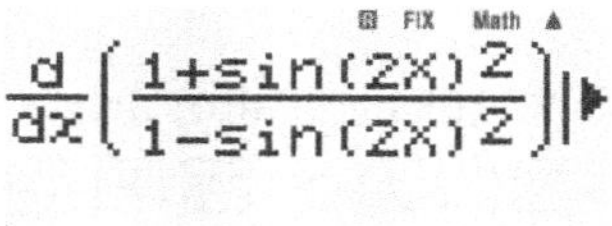

First value $x = -3.7$

[(−)] [3] [•] [7] [=]

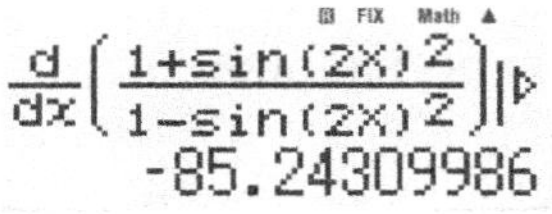

Which corresponds to the previous calculation.

Second value $x = 1.2$

[◄] [DEL] [DEL] [DEL] [DEL] [DEL]

[1] [•] [2] [=]

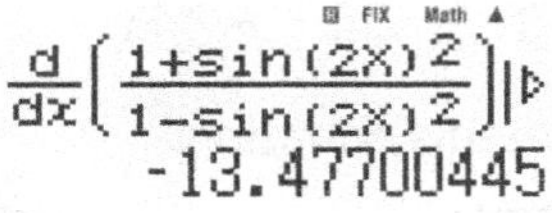

Which again corresponds to the previous calculation.

Finally the third value $x = 5.8$

[◄] [DEL] [DEL] [DEL] [DEL]

[5] [•] [8] [=]

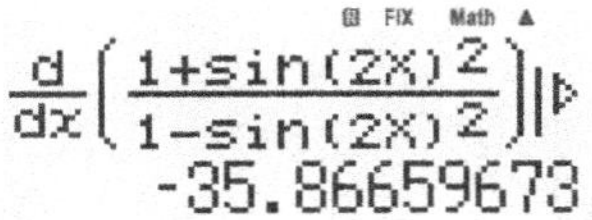

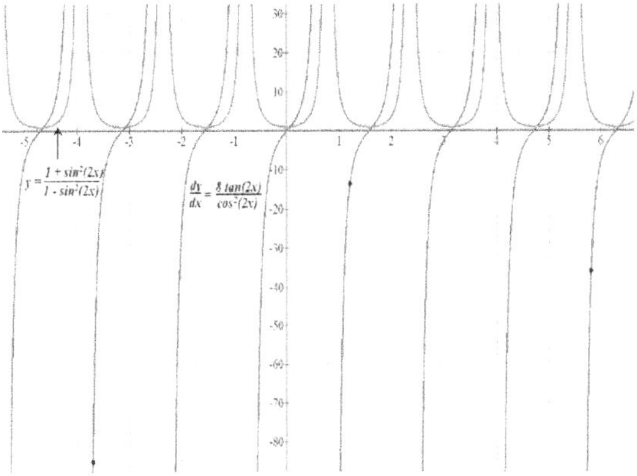

Compare with the previous calculation; these results affirm the correctness of the differentiation. Here is a plot of the function and its derivative. On the graph you can see the three test results on the derivative plot.

Example 5:

Find the stationary points of the following function and determine whether they are maximum or minimum.

$$y = x^4 - 8x^3 + 22x^2 - 24x + 12$$

This is a 4th order polynomial and the stationary points occur then $\frac{dy}{dx} = 0$.

Perform the differentiation,

$$\frac{dy}{dx} = 4x^3 - 24x^2 + 44x - 24$$

When $\frac{dy}{dx} = 0$ need to solve this equation

$$4x^3 - 24x^2 + 44x - 24 = 0$$

Divide by 4, $x^3 - 6x^2 + 11x - 6 = 0$

Solve this equation using your *fx-991ES* calculator. Select the equation mode (MODE 5).

```
1:COMP   2:CMPLX
3:STAT   4:BASE-N
5:EQN    6:MATRIX
7:TABLE  8:VECTOR
```

You are presented with a number of options, (4)

```
1:anX+bnY=cn
2:anX+bnY+cnZ=dn
3:aX2+bX+c=0
4:aX3+bX2+cX+d=0
```

Enter the coefficients of the equation,

1 = (−) 6 = 1 1 = (−) 6 =

=

=

=

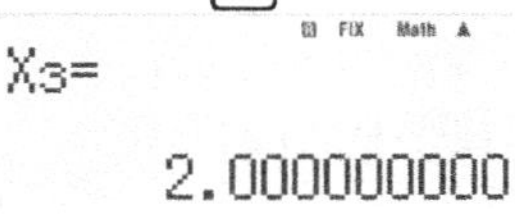

These are the values where the stationary points occur. To determine whether they are maximum or minimum use the differentiation feature on your calculator on the equation,

$$x^3 - 6x^2 + 11x - 6 = 0$$

Enter the keystrokes,

MODE 1 SHIFT ($\frac{d}{dx}$▪) ALPHA (X) SHIFT (x^3) − 6 RCL (X) x^2 + 1 1 RCL (X) − 6 ▶

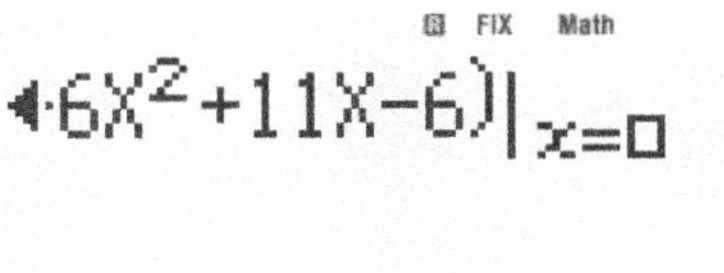

For the first point,

1 =

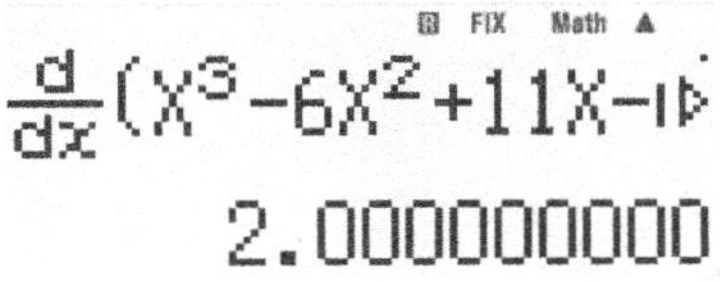

Positive value → minimum

For the second point,

◀ DEL DEL 3 =

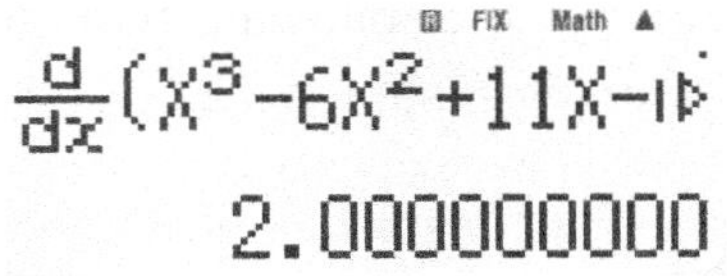

Positive value → minimum

For the third point,

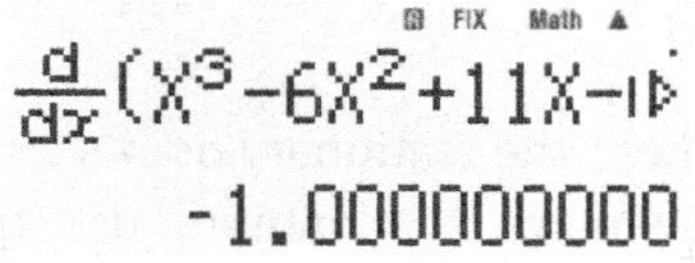

Negative value → maximum

A graph of the function and its derivative is show on the next page. You will observe the two minimums and one maximum occur when the derivative passes through the x-axis (when $y = 0$).

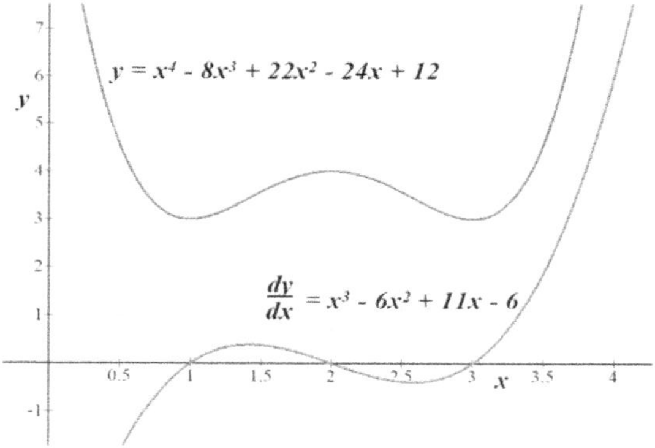

Example 6:

Find the stationary points of the following function and determine whether they are maximum or minimum.

$$y = (x-1)^3(x-2)^2$$

To find the stationary points, differentiate this equation,

$$\frac{dy}{dx} = (x-1)^3\frac{d}{dx}(x-2)^2 + (x-2)^2\frac{d}{dx}(x-1)^3$$

$$= (x-1)^3 2(x-2) + (x-2)^2 3(x-1)^2$$

$$\frac{dy}{dx} = (x-1)^2(x-2)[2(x-1)+3(x-2)]$$

$$\frac{dy}{dx} = (x-1)^2(x-2)[2x-2+3x-6]$$

$$= (x-1)^2(x-2)[5x-8]$$

This derivative has zero value when

$$x = 1, \qquad x = 2, \qquad x = 8/5$$

Need to determine what kind of stationary points these are, use your *fx-991ES* calculator.

The keystrokes are,

[SHIFT] ($\frac{d}{dx}$▪) [(] [ALPHA] (X) [−] [1] [)] [x^2] [(] [RCL] (X) [−] [2] [)] [(] [5] [RCL] (X) [−] [8] [)] [▶]

Point $x = 1$

[1] [=]

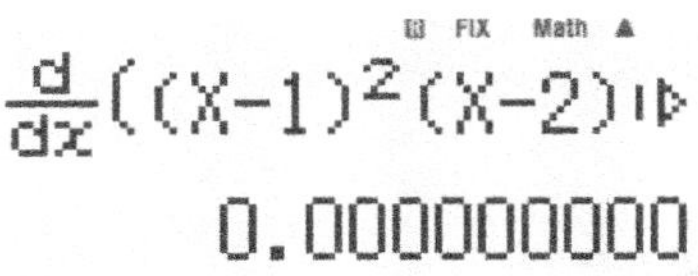

Zero → inflexion point

Point $x = 2$

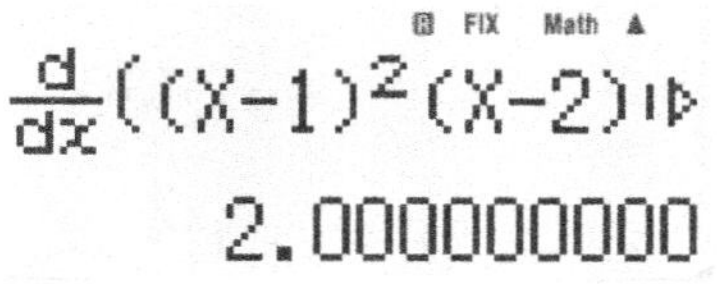

Positive → minimum

Point $x = 8/5$

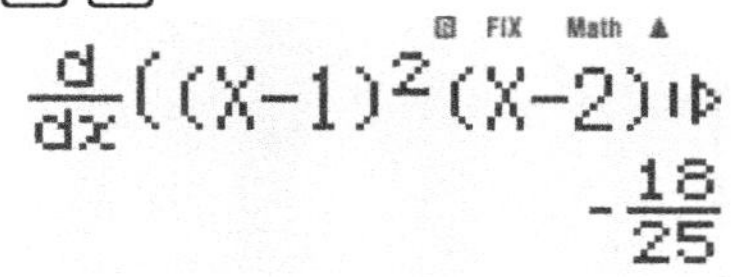

Negative → maximum

A graph of the function and its derivative is show here.

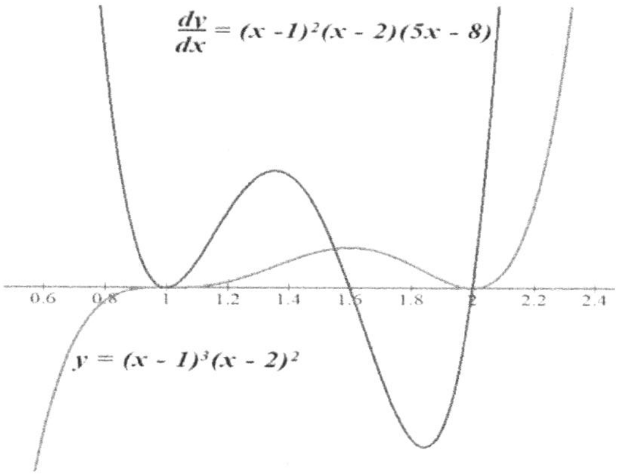

Note the saddle point at $x = 1$.

Example 7:

Find the stationary points of the following function and determine whether they are maximum or minimum.

$$y = \frac{5x^2}{(x+1)^4}$$

To find the stationary points, differentiate this equation,

$$\frac{dy}{dx} = \frac{5}{(x+1)^8}\left\{(x+1)^4\frac{d}{dx}x^2 - x^2\frac{d}{dx}(x+1)^4\right\}$$

$$= \frac{5}{(x+1)^8}\{(x+1)^4 2x - x^2 4(x+1)^3\}$$

$$= \frac{5(x+1)^3}{(x+1)^8}\{(x+1)\ \ 2x - x^2 4\}$$

$$= \frac{5}{(x+1)^5}\{2x(x+1) - 4x^2\}$$

$$= \frac{5}{(x+1)^5}\{2x^2 + 2x - 4x^2\}$$

$$= \frac{5}{(x+1)^5}\{-2x^2 + 2x\}$$

$$\frac{dy}{dx} = -\frac{10x(x-1)}{(x+1)^5}$$

As you can see there are two stationary points $x = 0$, $x = 1$. To determine what they are, use the differentiation feature on you *fx-991ES* calculator. The keystrokes are,

[SHIFT] ($\frac{d}{dx}$▪) [(−)] [▭/▭] [1] [0] [ALPHA] (X) [(] [RCL] (X) [−] [1] [)] [▼] [(] [RCL] (X) [+] [1] [)] [$x^{■}$] [5] [▶] [▶] [▶]

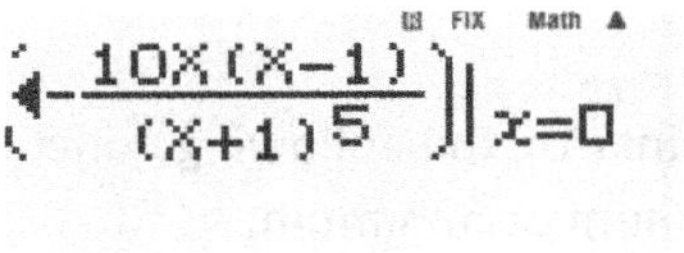

For value $x = 0$

[0] [=]

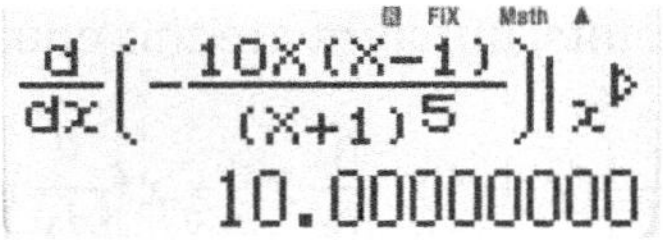

Positive number → minimum

For $x = 1$

[◀] [DEL] [DEL] [1] [=]

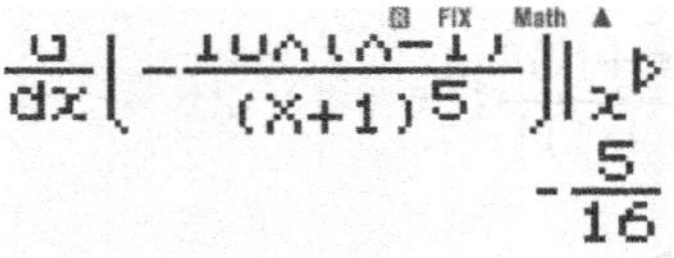

Negative number → a maximum

A graph of the function and its derivative is show here.

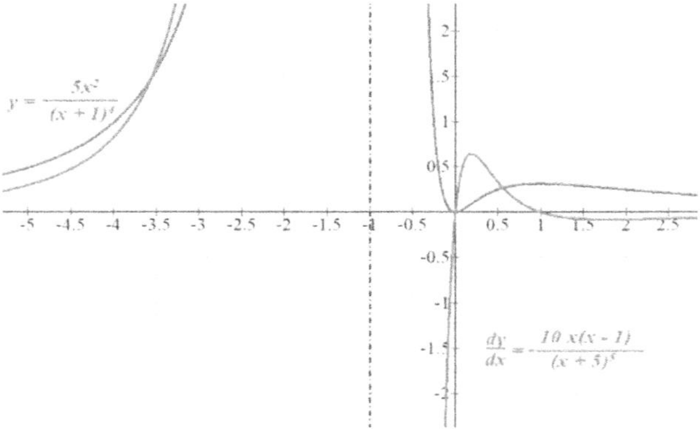

Note the location of the two turning points of the function at $x = 0$ and $x = 1$ where the derivative is zero.

Example 8:

Differentiate the following function,

$$y = \frac{3 - x}{\sqrt{3^2 - x^2}}$$

Use the quotient rule,

$$\frac{dy}{dx} = \frac{1}{(3^2 - x^2)}\left\{\sqrt{3^2 - x^2}\frac{d}{dx}(3 - x) - (3 - x)\frac{d}{dx}\sqrt{3^2 - x^2}\right\}$$

$$= \frac{1}{(3^2 - x^2)}\left\{-\sqrt{3^2 - x^2} - (3 - x)\frac{d}{dx}\sqrt{3^2 - x^2}\right\}$$

Let $X = \sqrt{3^2 - x^2}$ then $X^2 = 3^2 - x^2$, $\frac{d}{dx}X^2 = -2x$

$$2X\frac{dX}{dx} = -2x\,, \quad \frac{dX}{dx} = -\frac{x}{X}\,,$$

$$\frac{d}{dx}\sqrt{3^2 - x^2} = -\frac{x}{\sqrt{3^2 - x^2}}$$

$$\frac{dy}{dx} = \frac{1}{(3^2 - x^2)}\left\{-\sqrt{3^2 - x^2} + (3 - x)\frac{x}{\sqrt{3^2 - x^2}}\right\}$$

$$= \left\{-\frac{\sqrt{3^2 - x^2}}{(3^2 - x^2)} + \frac{(3 - x)}{(3^2 - x^2)}\frac{x}{\sqrt{3^2 - x^2}}\right\}$$

Consider,

$$\frac{\sqrt{u}}{u} = \frac{u^{\frac{1}{2}}}{u^{\frac{2}{2}}} = u^{\frac{1}{2}}\,u^{-\frac{2}{2}} = u^{\frac{1}{2}-\frac{2}{2}} = u^{-\frac{1}{2}} = \frac{1}{\sqrt{u}}$$

Therefore,

$$\frac{dy}{dx} = \left\{-\frac{1}{\sqrt{3^2 - x^2}} + \frac{(3 - x)}{(3^2 - x^2)}\frac{x}{\sqrt{3^2 - x^2}}\right\}$$

$$= \frac{1}{\sqrt{3^2 - x^2}}\left\{-1 + \frac{(3-x)x}{(3^2-x^2)}\right\}$$

$$= \frac{1}{\sqrt{3^2 - x^2}}\left\{\frac{-(3^2-x^2) + (3-x)x}{(3^2-x^2)}\right\}$$

$$= \frac{1}{\sqrt{3^2 - x^2}}\left\{\frac{-3^2+x^2 + 3x - x^2}{(3^2-x^2)}\right\}$$

$$= \frac{3}{\sqrt{3^2 - x^2}}\left\{\frac{-3+x}{(3^2-x^2)}\right\}$$

$$= -\frac{3}{\sqrt{3^2 - x^2}}\left\{\frac{3-x}{(3+x)(3-x)}\right\}$$

$$= -\frac{3}{\sqrt{3^2 - x^2}}\left\{\frac{1}{(3+x)}\right\} = -\frac{3}{(3+x)\sqrt{3^2 - x^2}}$$

$$= -\frac{3}{\sqrt{(3+x)^2(3^2-x^2)}}$$

Evaluate this derivative for

$x = -2.5, 1.6, 2.8$

Keystrokes for your *fx-991ES* calculator are,

[(−)] [3] [▭/▭] [√▭] [(] [3] [+] [ALPHA] (X) [)] [x^2] [(] [9] [−] [RCL] (X) [x^2] [)]

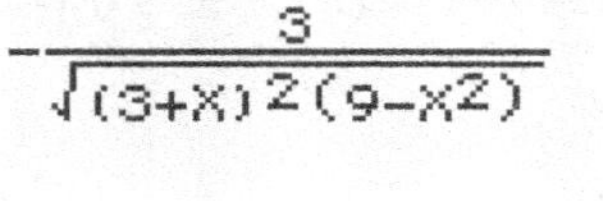

[CALC] X? [(−)] [2] [•] [5] [=] [S⇔D]

[CALC] X? [1] [•] [6] [=]

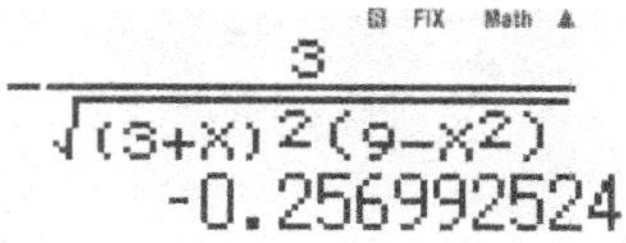

[CALC] X? [2] [•] [8] [=]

x	-2.5	1.6	2.8
dy/dx	-3.6181	-0.257	-0.4802

Test these results using the differentiation feature on your calculator. The original equation is,

$$y = \frac{3 - x}{\sqrt{3^2 - x^2}}$$

The keystrokes are,

[AC] [SHIFT] ($\frac{d}{dx}$■) [▭/▭] [3] [−] [ALPHA] (X) [▼] [√■] [9] [−]

[RCL] (X) [x^2] [▶] [▶] [▶]

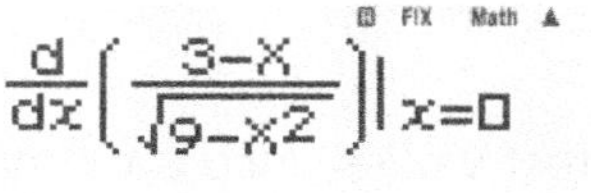

[(−)] [2] [•] [5] [=]

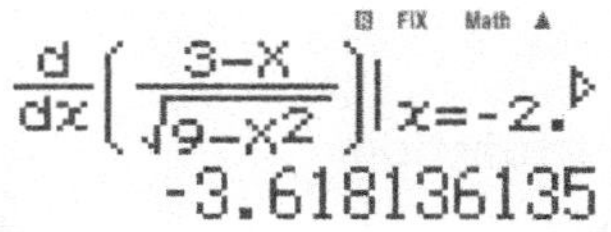

[◀] [DEL] [DEL] [DEL] [DEL] [DEL] [1] [•] [6] [=]

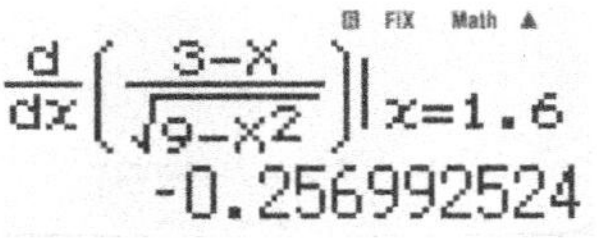

◀ DEL DEL DEL DEL
2 · 8 =

Compare these results with the calculations in the table and you will see a match, which confirms the differentiation is correct. A graph of the function and its derivative is shown below.

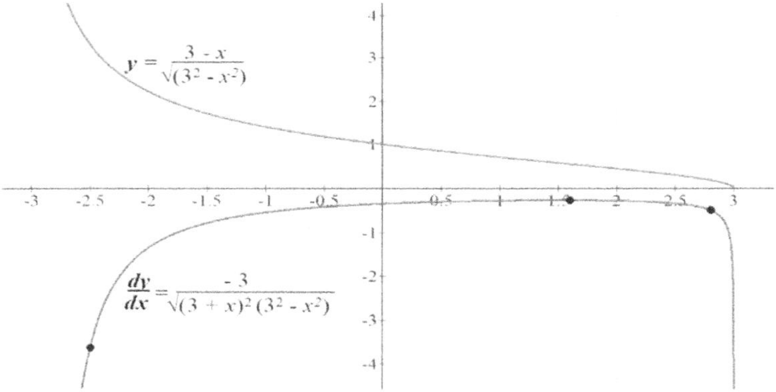

Note the location of the test values. Also the function does not exist beyond $y > 3$.

Example 9:

Differentiate the following function,

$$y = \frac{1-2x}{(1+3x)^2}$$

Use the quotient rule,

$$\frac{dy}{dx} = \frac{1}{(1+3x)^4}\left\{(1+3x)^2\frac{d}{dx}(1-2x) - (1-2x)\frac{d}{dx}(1+3x)^2\right\}$$

$$= \frac{1}{(1+3x)^4}\left\{-2(1+3x)^2 - (1-2x)\frac{d}{dx}(1+3x)^2\right\}$$

Let $X = 1 + 3x$, $\frac{dX}{dx} = 3,$

$$\frac{d}{dx}X^2 = 2X\frac{dX}{dx} = 2(1+3x)3$$

$$\frac{dy}{dx} = \frac{1}{(1+3x)^4}\{-2(1+3x)^2 - (1-2x)6(1+3x)\}$$

$$= \frac{1}{(1+3x)^3}\{-2(1+3x) \quad - (1-2x)6\}$$

$$= \frac{1}{(1+3x)^3}\{-2 - 6x - 6 + 12x\}$$

$$= \frac{1}{(1+3x)^3}\{6x - 8\}$$

Leaving

$$\frac{dy}{dx} = \frac{(6x-8)}{(1+3x)^3}$$

Evaluate this derivative for

$x = -1.5, \; -1.0, \; 0.4$

Keystrokes for your *fx-991ES* calculator are,

[fraction] (6 ALPHA (X) − 8) ▼ (1 + 3 RCL (X)) SHIFT (x^3)

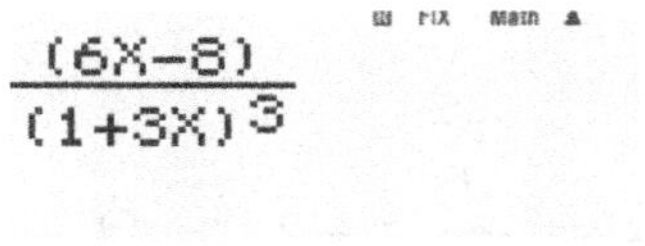

[CALC] X? [(−)] [1] [•] [5] [=] [S⇔D]

[CALC] X? [(−)] [1] [=] [S⇔D]

[CALC] X? [0] [•] [4] [=] [S⇔D]

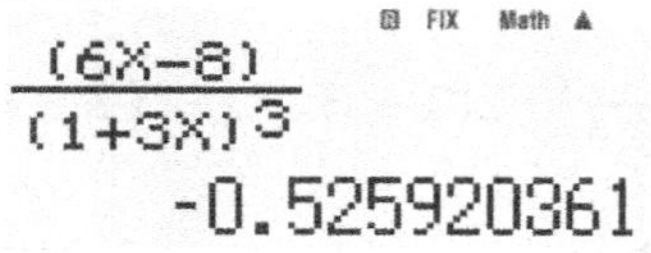

x	-1.5	-1.0	0.4
dy/dx	0.396	1.75	-0.526

Test these results using the differentiation feature on your *fx-991ES* calculator. The original equation is,

$$y = \frac{1 - 2x}{(1 + 3x)^2}$$

The keystrokes are,

[AC] [SHIFT] ($\frac{d}{dx}$■) [▭/▭] [1] [−] [2] [ALPHA] (X) [▼] [(] [1] [+] [3] [RCL] (X) [)] [x^2]

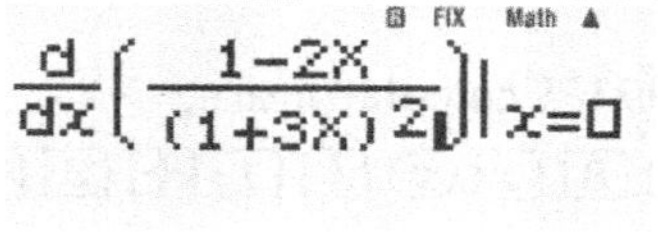

▶ ▶ (−) 1 • 5 =

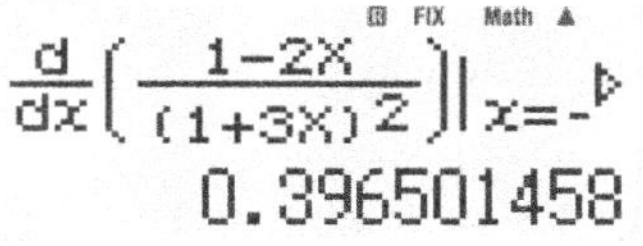

◀ DEL DEL DEL = S⇔D

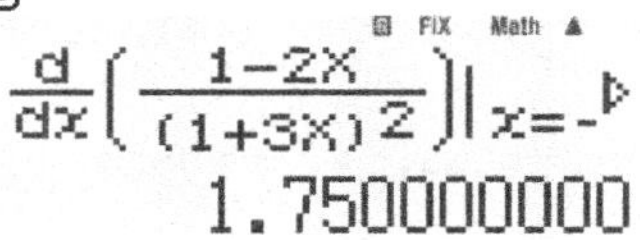

◀ DEL DEL DEL
0 • 4 = S⇔D

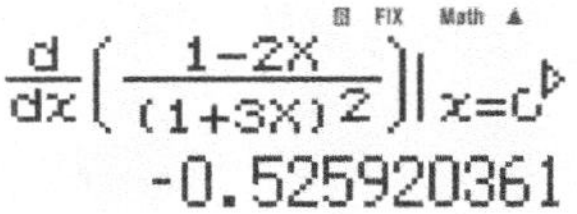

Compare with the results in the table which confirm the differentiation is correct. A plot of the function and its derivative is shown here,

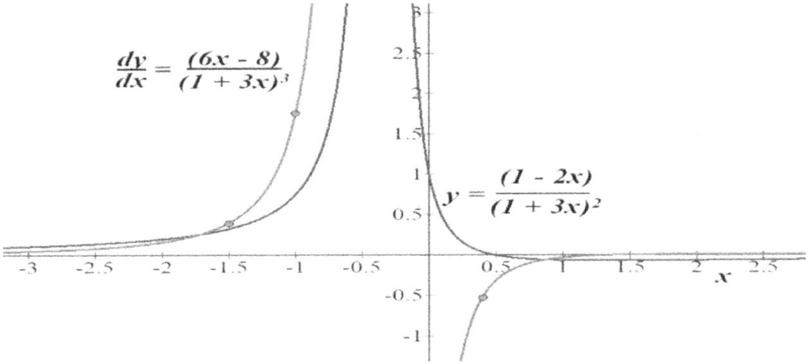

You will note the three test values and the presence of the asymptote at $x = -\frac{1}{3}$.

Example 10:

Differentiate the following function,

$$y^2 = x^3 \cos(3x)$$

Use the product rule,

$$\frac{dy^2}{dx} = x^3 \frac{d}{dx}\cos(3x) + \cos(3x)\frac{d}{dx}x^3$$

$$2y\frac{dy}{dx} = x^3 \frac{d}{dx}\cos(3x) + \cos(3x)\frac{d}{dx}x^3$$

$$2y\frac{dy}{dx} = -x^3 3\sin(3x) + \cos(3x)\frac{d}{dx}x^3$$

$$2y\frac{dy}{dx} = -x^3 3\sin(3x) + \cos(3x)\,3x^2$$

$$2y\frac{dy}{dx} = 3x^2[\cos(3x) - x\sin(3x)]$$

$$y\frac{dy}{dx} = \frac{3}{2}x^2[\cos(3x) - x\sin(3x)]$$

Divide both sides by y,

$$\frac{dy}{dx} = \frac{3}{2}\frac{x^2[\cos(3x) - x\sin(3x)]}{y}$$

The original equation was,

$$y^2 = x^3\cos(3x), \quad y = \sqrt{x^3\cos(3x)}$$

$$\frac{dy}{dx} = \frac{3}{2}\frac{x^2[\cos(3x) - x\sin(3x)]}{\sqrt{x^3\cos(3x)}}$$

Evaluate this derivative for $x = -1.3,\ 1.6,\ 2.4$.

The calculator must be configured in its radian mode ([SHIFT] (SETUP) [4]); the keystrokes for your *fx-991ES* calculator are

[■/□] [3] [ALPHA] (X) [x^2] [(] [cos] [3] [RCL] (X) [)] [−] [RCL] (X) [sin] [3] [RCL] (X) [)]
[)] [▼] [2] [√■] [RCL] (X) [SHIFT] (x^3) [cos] [3] [RCL] (X) [)]

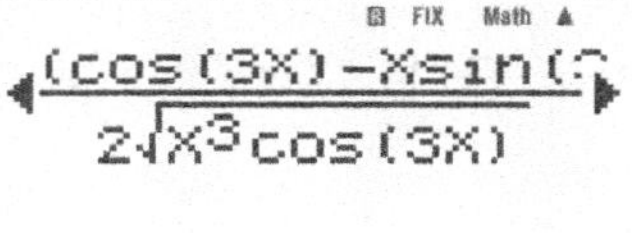

[CALC] X? [(−)] [1] [•] [3] [=]

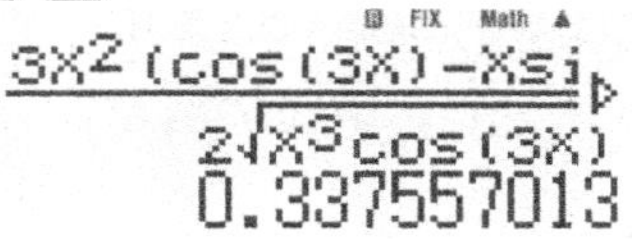

[CALC] X? [1] [•] [6] [=]

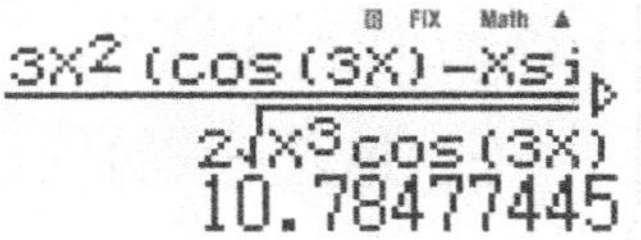

[CALC] X? [2] [•] [4] [=]

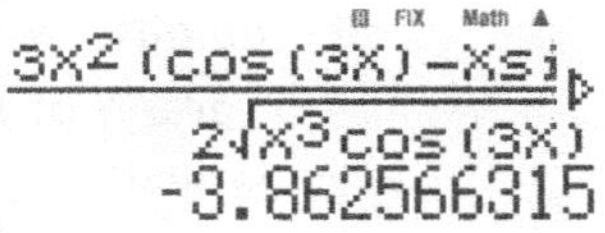

x	-1.3	1.6	2.4
dy/dx	0.3375	10.7847	-3.862

Test these results using the differentiation feature on your calculator.
The original equation is,

$$y = \sqrt{x^3 \cos(3x)}$$

The keystrokes are,

[AC] [SHIFT] ($\frac{d}{dx}$■) [√■] [ALPHA] (X) [SHIFT] (x^3) [cos] [3] [RCL] (X) [)] [▶] [▶] [(−)] [1] [•] [3] [=]

[◀] [DEL] [DEL] [DEL] [DEL] [DEL]
[1] [•] [6] [=]

[◀] [DEL] [DEL] [DEL] [DEL]
[2] [•] [4] [=]

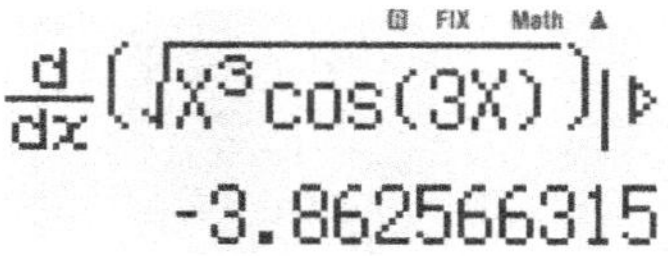

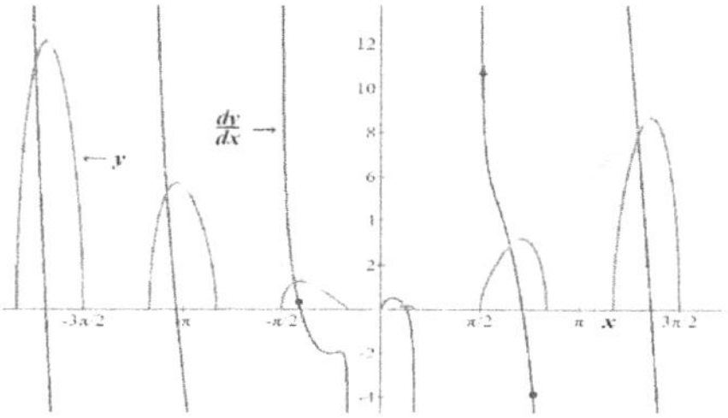

Compare with the calculations in the table, the correctness of the differentiation is confirmed. A graph of the function and its derivative are shown here. There are several interesting features, notice there are gaps in the function. These occur when the value within the square root are negative. Observe also the increase in the height of the lobes follow a x^2 curve as shown in green. The maxima of the function occur when the derivative passes through the x-axis.

Example 11:

Differentiate the following function,

$$(y-3)^2+(x-2)^2=5$$

$$\frac{d}{dx}(y-3)^2+\frac{d}{dx}(x-2)^2=0$$

$$\frac{d}{dx}(y^2-6y+9) \quad +\frac{d}{dx}(x^2-4x+2) \quad =0$$

$$2y\frac{dy}{dx}-6\frac{dy}{dx}+2x-4=0$$

$$2y\frac{dy}{dx}-6\frac{dy}{dx}=4-2x$$

$$\frac{dy}{dx}(2y-6)=4-2x$$

$$\frac{dy}{dx}(y-3)=2-x$$

$$\frac{dy}{dx}=\frac{2-x}{y-3}$$

The original equation is,

$$(y-3)^2+(x-2)^2=5$$

$$(y-3)^2=5-(x-2)^2$$

$$y-3=\pm\sqrt{5-(x-2)^2}$$

$$y - 3 = \pm\sqrt{5 - x^2 + 4x - 4}y - 3 = \pm\sqrt{1 - x^2 + 4x}$$

The derivative becomes,

$$\frac{dy}{dx} = \frac{2 - x}{\pm\sqrt{1 - x^2 + 4x}}$$

Evaluate this derivative (for the + value only) for $x = -0.2,\ 1.0,\ 3.5.$

The keystrokes for your *fx-991ES* calculator are,

[▭/▭] [2] [−] [ALPHA](X) [▼] [√▭] [1] [−] [RCL](X) [x^2] [+] [4] [RCL](X)

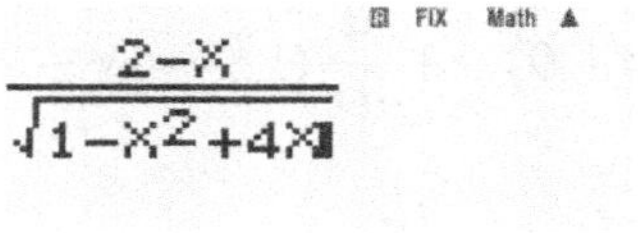

[CALC] X? [(−)] [0] [•] [2] [=] [S⇔D]

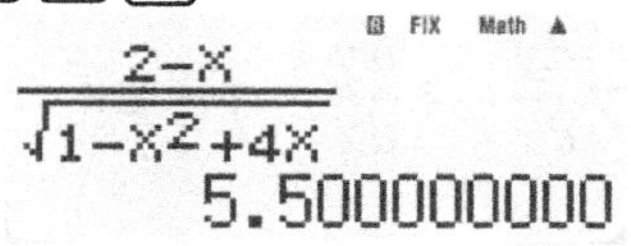

[CALC] X? [1] [•] [0] [=] [S⇔D]

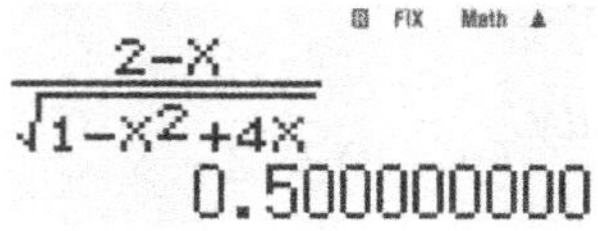

[CALC] X? [3] [•] [5] [=] [S⇔D]

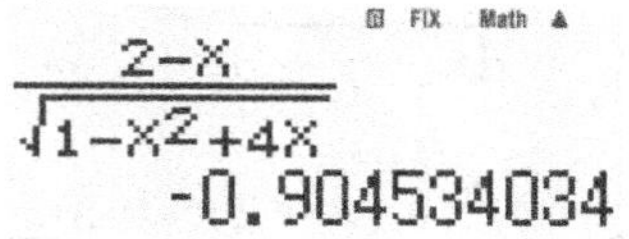

x	-0.2	1.0	3.5
dx/dy	5.5	0.5	-0.9045

Test these results using the differentiation feature on your calculator. The original equation is,

$$y = 3 + \sqrt{1 - x^2 + 4x}$$

The keystrokes are,

[AC] [SHIFT] ($\frac{d}{dx}\blacksquare$) [3] [+] [√■] [1] [−] [ALPHA] (X) [x^2] [+] [4] [RCL] (X)

$\frac{d}{dx}\left(3+\sqrt{1-X^2+4X}\right)|$

[▶] [▶] [(−)] [0] [•] [2] [=]

$\frac{d}{dx}\left(3+\sqrt{1-X^2+4X}\right)|$

5.500000000

[◀] [DEL] [DEL] [DEL] [DEL] [DEL]
[1] [•] [0] [=]

$\frac{d}{dx}\left(3+\sqrt{1-X^2+4X}\right)|$

0.500000000

[◀] [DEL] [DEL] [DEL] [DEL]
[3] [•] [5] [=]

$\frac{d}{dx}\left(3+\sqrt{1-X^2+4X}\right)|$

-0.904534034

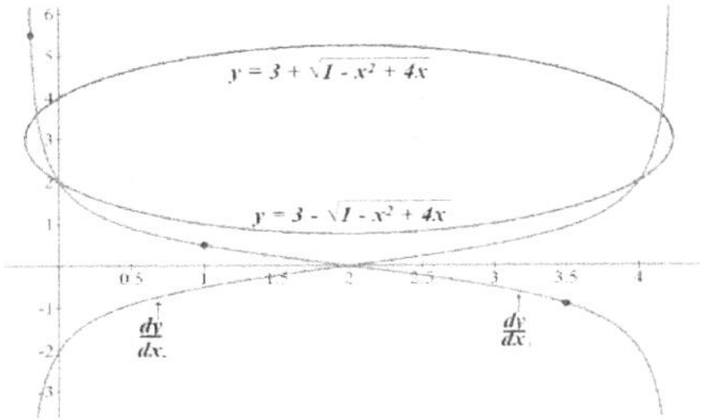

Compare with the results in the table, this proves the correctness of the differentiation. A plot of the function and its derivative is shown below. Also included in this graph is the negative value as shown in red. The corresponding derivative curves are also shown. Notice the value of the derivatives at the maximum and minimum values and the asymptotes.

Example 12:

Differentiate the following function,

$$y^2 + 2x^4 - 2x = 6$$

$$\frac{d}{dx}y^2 + \frac{d}{dx}2x^4 - \frac{d}{dx}2x = 0$$

$$2y\frac{dy}{dx} + 8x^3 - 2 = 0$$

$$y\frac{dy}{dx} + 4x^3 - 1 = 0$$

$$y\frac{dy}{dx} = 1 - 4x^3$$

$$\frac{dy}{dx} = \frac{1 - 4x^3}{y}$$

The original equation,

$$y^2 + 2x^4 - 2x = 6$$

$$y^2 = 6 - 2x^4 + 2x$$

$$y = \sqrt{6 - 2x^4 + 2x}$$

Leaving,

$$\frac{dy}{dx} = \frac{1 - 4x^3}{\sqrt{6 - 2x^4 + 2x}}$$

$$\frac{dy}{dx} = \frac{1 - 4x^3}{\sqrt{6 + 2x(1 - x^3)}}$$

Evaluate this derivative for $x = -1,\ 0.5$ and 1.25. The keystrokes for your *fx-991ES* calculator are,

[▭/▭] [1] [−] [4] [ALPHA] (X) [SHIFT] (x^3) [▼] [√■] [6] [+] [2] [RCL] (X) [(] [1] [−] [RCL] (X) [SHIFT] (x^3) [)]

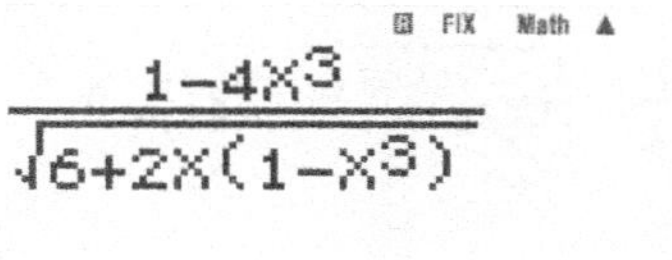

Evaluate this derivative for $x = -1,\;\; 0.5$ and 1.25

[CALC] X? [(−)] [1] [=] [S⇔D]

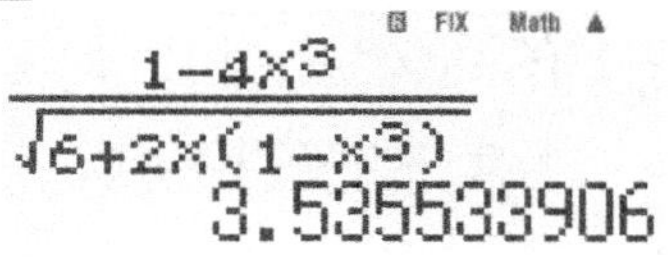

[CALC] X? [0] [•] [5] [=] [S⇔D]

[CALC] X? [1] [•] [2] [5] [=] [S⇔D]

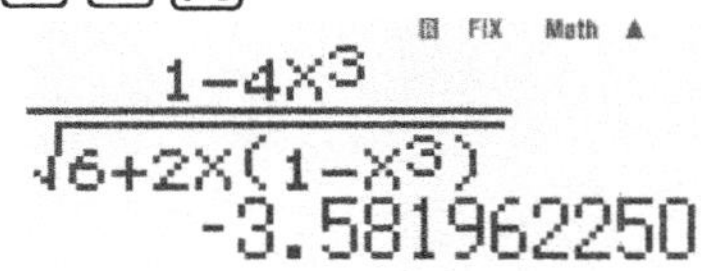

x	-1	0.5	1.25
dy/dx	3.535	0.1906	-3.5819

Test these results using the differentiation feature on your *fx-991ES* calculator. The original equation is,

$$y = \sqrt{6 - 2x^4 + 2x}$$

The keystrokes are,

[AC] [SHIFT] ($\frac{d}{dx}\blacksquare$) [$\sqrt{\blacksquare}$] [6] [−] [2] [ALPHA] (X) [$x^{\blacksquare}$] [4] [▶] [+] [2] [RCL] (X)

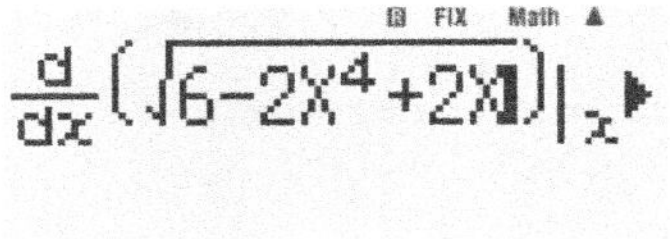

[▶] [▶] [(−)] [1] [=]

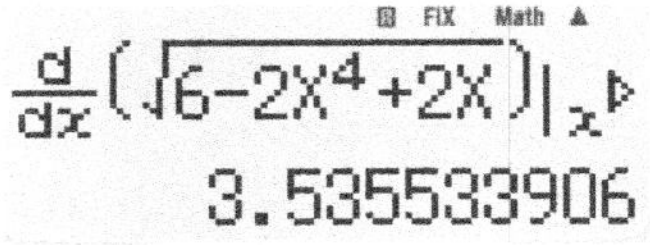

[◀] [DEL] [DEL] [DEL]
[0] [•] [5] [=]

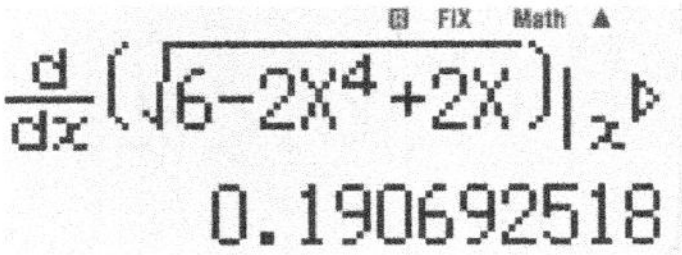

[◀] [DEL] [DEL] [DEL] [DEL]
[1] [•] [2] [5] [=]

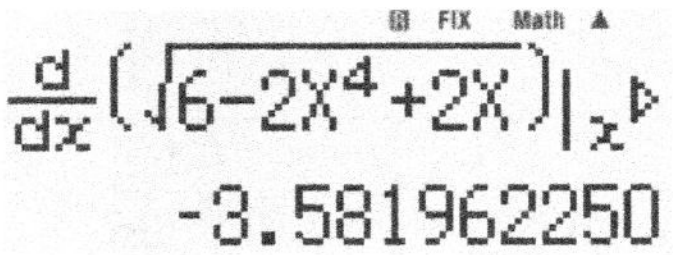

Compare with the calculations in the table, these results confirm the correctness of the differentiation. A plot of the function and its derivative is shown on the next page.

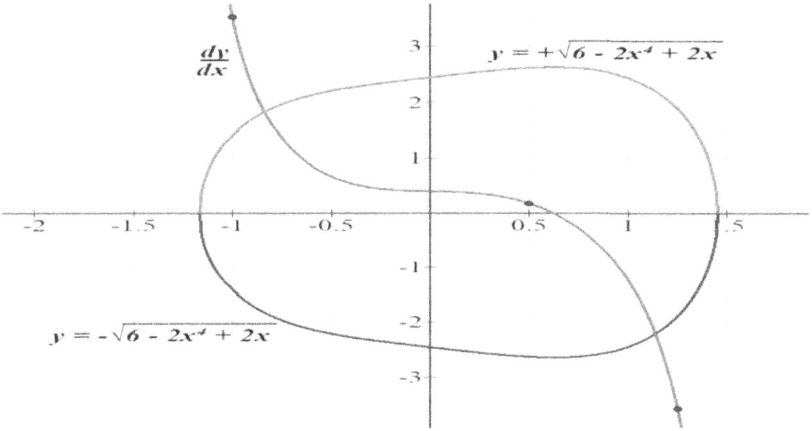

You will observe the positive and negative functions are shown here and it does not exist beyond the range (when the value under the square root becomes negative). The three test values are shown on the derivative plot which passes through the x-axis when the maximum and minimum occur.

2 Integration Examples

In this chapter there are sixteen worked examples of integrations. Use your fx-991ES calculator to confirm the correctness of the results.

Example 1:

Perform the integration

$$\int_{0}^{\pi/3} \frac{1 - \tan(x)}{e^{2x}\cos^2(x) + 1} dx$$

Let $u = e^{2x}\cos^2(x)$

Differentiate,

$$\frac{du}{dx} = \frac{d}{dx}[e^{2x}\cos^2(x)]$$

$$= \cos^2(x)\frac{d}{dx}e^{2x} + e^{2x}\frac{d}{dx}\cos^2(x)$$

$$= 2e^{2x}\cos^2(x) - 2\sin(x)\cos(x)e^{2x}$$

$$= 2e^{2x}\cos^2(x) - 2\sin(x)\cos(x)e^{2x}\frac{\cos(x)}{\cos(x)}$$

$$= 2e^{2x}\cos^2(x) - 2\cos^2(x)e^{2x}\frac{\sin(x)}{\cos(x)}$$

$$= 2e^{2x}\cos^2(x) - 2\cos^2(x)e^{2x}\tan(x)$$

Since $\qquad u = e^{2x}\cos^2(x)$

$$\frac{du}{dx} = 2u - 2u\tan(x) = 2u\,[1 - \tan(x)]$$

$$\frac{du}{1-\tan(x)} = dx$$

The integral is

$$\int_0^{\pi/3} \frac{1-\tan(x)}{e^{2x}\cos^2(x)+1}dx$$

$$\int_0^{\pi/3} \frac{1-\tan(x)}{u+1}\cdot\frac{du}{2u[1-\tan(x)]}$$

$$= \frac{1}{2}\int_0^{\pi/3} \frac{du}{u(u+1)}$$

To perform this integration, need to express it as *partial fractions*,

$$\frac{1}{u(u+1)} = \frac{A}{u} + \frac{B}{u+1}$$

$$1 = A(u+1) + Bu$$

or

$$1 = Au + A + Bu$$

Equate coefficients,

$$1 = A$$

$$0 = A + B$$

$$B = -1$$

$$\frac{1}{u(u+1)} = \frac{A}{u} + \frac{B}{u+1} = \frac{1}{u} - \frac{1}{u+1}$$

$$\frac{1}{2}\int_0^{\pi/3} \frac{du}{u(u+1)} = \frac{1}{2}\int_0^{\pi} \left[\frac{1}{u} - \frac{1}{u+1}\right] du$$

Perform the integration,

$$\frac{1}{2}\int_0^{\pi} \left[\frac{1}{u} - \frac{1}{u+1}\right] du$$

$$= \frac{1}{2}\left[\ln(u) - \ln(u+1)\right]_0^{\pi/3}$$

$$= \frac{1}{2}\left[\ln\left(\frac{u}{u+1}\right)\right]_0^{\pi/3}$$

Since $u = e^{2x}\cos^2(x)$

$$= \frac{1}{2}\left[\ln\left(\frac{e^{2x}\cos^2(x)}{e^{2x}\cos^2(x)+1}\right)\right]_0^{\pi/3}$$

$$= \frac{1}{2}\left[\ln\left(\frac{e^{2\frac{\pi}{3}}\cos^2(\frac{\pi}{3})}{e^{2\frac{\pi}{3}}\cos^2\left(\frac{\pi}{3}\right)+1}\right) - \ln\left(\frac{e^0\cos^2(0)}{e^0\cos^2(0)+1}\right)\right]$$

$$= \frac{1}{2}\left[\ln\left(\frac{e^{2\frac{\pi}{3}}\cos^2(\frac{\pi}{3})}{e^{2\frac{\pi}{3}}\cos^2\left(\frac{\pi}{3}\right)+1}\right) - \ln\left(\frac{1}{2}\right)\right]$$

Perform this calculation on your *fx-991ES* calculator, ensure it is in radian mode ([SHIFT] [MODE] [4]). The keystrokes for your *fx*-991ES; this is a two stage calculation, first calculate $e^{2\frac{\pi}{3}}\cos^2(\frac{\pi}{3})$ this gives [Ans].

[SHIFT] ($e^{\blacksquare}$) [2] [SHIFT] (π) [$\frac{\blacksquare}{\square}$] [3] [▶] [▶] [cos] [SHIFT] (π) [$\frac{\blacksquare}{\square}$] [3] [▶] [)] [$x^2$] [ALPHA] (:)

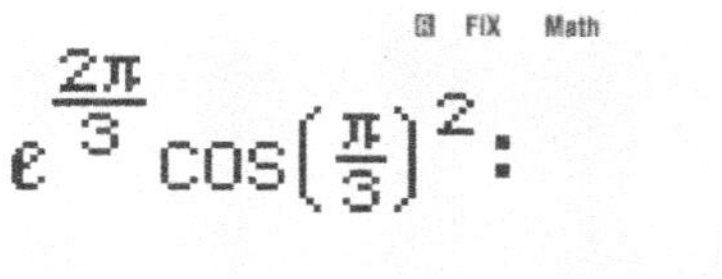

Which now feeds into the second Stage.

$$= \frac{1}{2}\left[\ln\left(\frac{\text{Ans}}{\text{Ans}+1}\right) - \ln\left(\frac{1}{2}\right)\right]$$

[1] [▭/▭] [2] [▶] [(] [ln] [Ans] [▭/▭] [Ans] [+] [1] [▶] [)] [−] [ln] [1] [▭/▭] [2]
[▶] [)] [)] [=] [=]

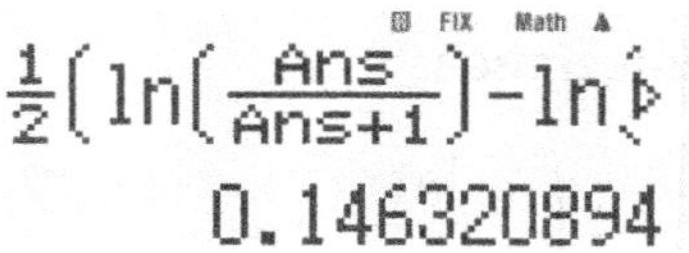

Confirm this result, consider the original integral

$$\int_0^{\pi/3} \frac{1-\tan(x)}{e^{2x}\cos^2(x)+1}\,dx$$

The keystrokes to perform this Integration are:

[∫▭] [▭/▭] [1] [−] [tan] [ALPHA] (X) [)] [▼] [SHIFT] ($e^{\blacksquare}$) [2] [RCL] (X) [▶] [cos] [RCL] (X)
[)] [x^2] [+] [1] [▼] [0] [▲] [SHIFT] (π) [÷] [3] [=]

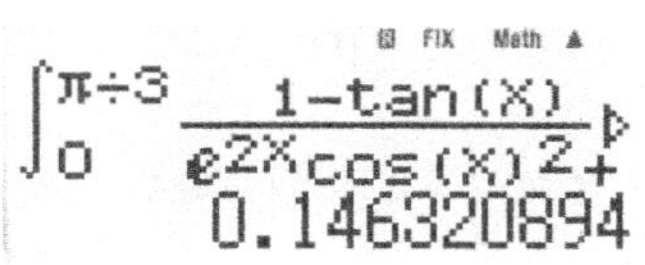

As you can see the two calculations produce the same result which suggests the integration is correct. A graph of the function is shown on the next page.

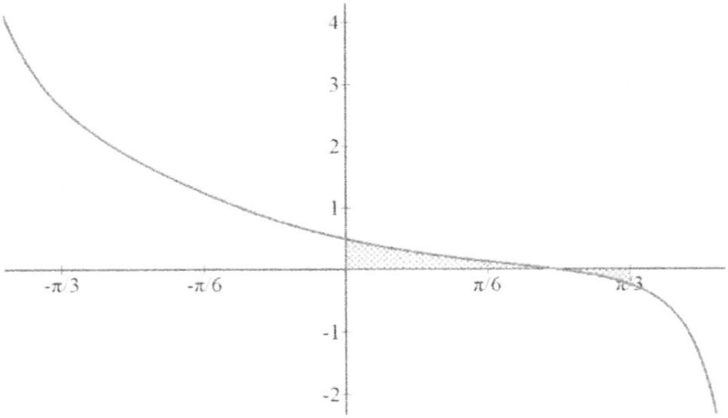

Example 2:

Perform the following integration,

$$\int_1^2 \frac{\ln(x)}{(x-3)^2}dx$$

This integration can be solved by using integration by parts which involves the variables u and v so that,

$$\int_1^2 \frac{\ln(x)}{(x-3)^2} = \int v\frac{du}{dx}dx = uv - \int u\frac{dv}{dx}dx$$

then

$$v = \ln(x) \text{ and } \frac{du}{dx} = (x-3)^{-2}$$

$$\frac{dv}{dx} = \frac{1}{x} \text{ and } u = \int (x-3)^{-2}dx = -\frac{1}{(x-3)}$$

$$\int_1^2 \frac{\ln(x)}{(x-3)^2}dx = -\left[\frac{\ln(x)}{(x-3)}\right]_1^2 + \int_1^2 \frac{dx}{x(x-3)}$$

Partial fractions,

$$\frac{1}{x(x-3)} = \frac{A}{x} + \frac{B}{x-3}$$

$$1 = A(x-3) + Bx$$

Expand

$$1 = Ax - 3A + Bx$$

Equate coefficients,

$$1 = -3A \text{ and } 0 = A + B, \quad A = -\frac{1}{3} \text{ and } B = \frac{1}{3}$$

Therefore,

$$\frac{1}{x(x-3)} = -\frac{1}{3x} + \frac{1}{3(x-3)}$$

The integration becomes,

$$\int_1^2 \frac{\ln(x)}{(x-3)^2}dx = -\left[\frac{\ln(x)}{(x-3)}\right]_1^2 + \frac{1}{3}\int_1^2 \frac{dx}{(x-3)} - \frac{1}{3}\int_1^2 \frac{dx}{x}$$

$$= -\left[\frac{\ln(x)}{(x-3)}\right]_1^2 + \frac{1}{3}[\ln(|x-3|)]_1^2 - \frac{1}{3}[\ln(|x|)]_1^2$$

$$= -\left[\frac{\ln(x)}{(x-3)}\right]_1^2 + \frac{1}{3}\left[\ln\left(\frac{|x-3|}{|x|}\right)\right]_1^2$$

$$= -\frac{\ln(2)}{(2-3)} + \frac{\ln(1)}{(1-3)} + \frac{1}{3}\ln\left(\frac{|2-3|}{|2|}\right) - \frac{1}{3}\ln\left(\frac{|1-3|}{|1|}\right)$$

$$= \frac{\ln(2)}{1} - \frac{\ln(1)}{2} + \frac{1}{3}\ln\left(\frac{1}{2}\right) - \frac{1}{3}\ln\left(\frac{2}{1}\right)$$

$$= \ln(2) - \frac{1}{3}\ln(2) - \frac{1}{3}\ln(2) = \frac{\ln(2)}{3}$$

$$\int_1^2 \frac{\ln(x)}{(x-3)^2}\,dx = \frac{\ln(2)}{3}$$

The keystrokes for this calculation,

[ln] [2] [)] [fraction] [3] [=]

The original integration is,

$$\int_1^2 \frac{\ln(x)}{(x-3)^2}\,dx$$

Verify the result by using the integration feature on your *fx* calculator,

[∫] [fraction] [ln] [ALPHA] (X) [)] [▼] [(] [RCL] (X) [−] [3] [)] [x^2] [▼] [1] [▲] [2] [=]

Which confirms the integration is correct. Here is a plot of the function and the area of the integration,

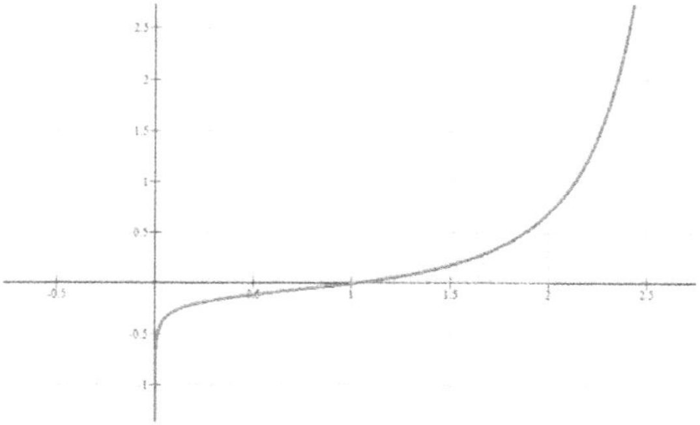

Example 3:

Perform the following integration,

$$\int_0^2 \frac{dx}{x^2 + 4x + 29}$$

Whenever you see an integral with a quadratic expression in the denominator then you need to express it in terms of *sum of the squares*.

Let,

$$x^2 + 4x + 29 = (x + A)^2 + B^2$$

Expand the brackets,

$$x^2 + 4x + 29 = x^2 + 2xA + A^2 + B^2$$

Equate coefficients,

$$4 = 2A$$
$$29 = A^2 + B^2$$
$$A = 2$$
$$29 = 4 + B^2$$
$$B^2 = 29 - 4 = 25$$

The denominator becomes,

$$x^2 + 4x + 29 = (x + A)^2 + B^2$$

$$= (x + 2)^2 + 25$$

The integral becomes,

$$\int_0^2 \frac{dx}{x^2 + 4x + 29} = \int_0^2 \frac{dx}{(x + 2)^2 + 25}$$

Let $u = x + 2, \quad du = dx$

$$\int_0^2 \frac{dx}{(x+2)^2 + 25} = \int_0^2 \frac{du}{u^2 + 5^2}$$

Using the well known result,

$$\int \frac{du}{u^2 + a^2} = \frac{1}{a}\tan^{-1}\left(\frac{u}{a}\right)$$

$$\int_0^2 \frac{du}{u^2 + 5^2} = \left[\frac{1}{5}\tan^{-1}\left(\frac{u}{5}\right)\right]_0^2$$

But $u = x + 2$

$$\int_0^2 \frac{dx}{(x+2)^2 + 25} = \left[\frac{1}{5}\tan^{-1}\left(\frac{x+2}{5}\right)\right]_0^2$$

$$= \frac{1}{5}\left[\tan^{-1}\left(\frac{2+2}{5}\right) - \tan^{-1}\left(\frac{2}{5}\right)\right]$$

Need to evaluate,

$$\frac{1}{5}\left[\tan^{-1}\left(\frac{4}{5}\right) - \tan^{-1}\left(\frac{2}{5}\right)\right]$$

Put your *fx-991ES* in radian mode ([SHIFT] [MODE] [4]). The keystrokes for the calculation are,

[1] [▭/▭] [5] [▶] [(] [SHIFT] $(\tan^{-1})$ [4] [▭/▭] [5] [▶] [)] [−] [SHIFT] $(\tan^{-1})$ [2] [▭/▭] [5] [▶] [)] [)] [=]

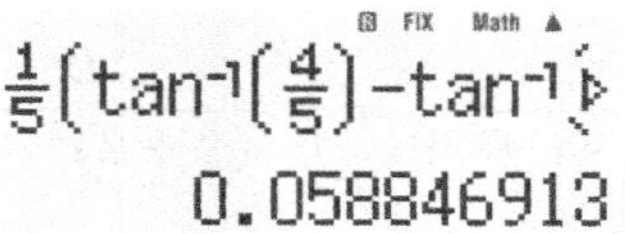

Verify this result by performing the Integral directly on your calculator,

$$\int_0^2 \frac{dx}{x^2 + 4x + 29}$$

The keystrokes for your calculator are,

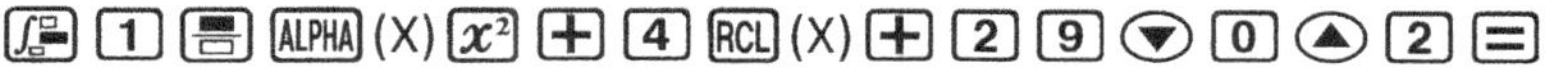

Here is a plot of the function and the integrated area.

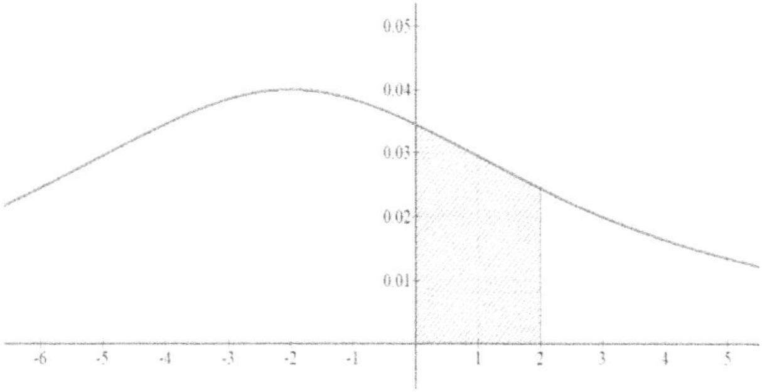

Example 4:

Perform the following integration,

$$\int_1^2 \frac{dx}{4x^2 - 12x + 5}$$

Again whenever you see an integral with a quadratic expression in the denominator then you need to express it in terms of *sum of the squares*. First take the 4 outside the integral.

$$\int_1^2 \frac{dx}{4x^2 - 12x + 5} = \frac{1}{4}\int_1^2 \frac{dx}{x^2 - 3x + \frac{5}{4}}$$

Let,

$$x^2 - 3x + \frac{5}{4} = (x + A)^2 + B^2$$

Expand the right hand side,

$$x^2 - 3x + \frac{5}{4} = x^2 + 2xA + A^2 + B^2$$

Equate coefficients,

$$-3 = 2A$$

$$\frac{5}{4} = A^2 + B^2$$

$$A = -\frac{3}{2}, \quad \frac{5}{4} = \frac{9}{4} + B^2, \qquad B^2 = \frac{5}{4} - \frac{9}{4} = -1$$

The integral

$$\frac{1}{4}\int_1^2 \frac{dx}{x^2 - 3x + \frac{5}{4}} = \frac{1}{4}\int_1^2 \frac{dx}{(x + A)^2 + B^2}$$

$$A = -\frac{3}{2} \text{ and } B^2 = -1$$

$$= \frac{1}{4}\int_1^3 \frac{dx}{\left(x - \frac{3}{2}\right)^2 - 1}$$

$$\frac{1}{4}\int_1^2 \frac{dx}{\left(x - \frac{3}{2}\right)^2 - 1} = \frac{1}{4}\int_1^2 \frac{dx}{\frac{1}{4}(2x - 3)^2 - \frac{4}{4}}$$

$$= \int_1^2 \frac{dx}{(2x-3)^2 - 4}$$

Let $u = 2x - 3, \;\; du = 2\,dx$

$$\frac{1}{2}\int_1^2 \frac{du}{u^2 - 2^2}$$

Express the integral as,

$$\frac{1}{2}\int_1^2 \frac{du}{(u+2)(u-2)}$$

Express the denominator as partial fractions,

$$\frac{1}{(u+2)(u-2)} = \frac{A}{u+2} + \frac{B}{u-2}$$

$$1 = A(u-2) + B(u+2)$$

$$1 = Au - 2A + Bu + 2B$$

Equate coefficients,

$$1 = -2A + 2B$$
$$0 = A + B, \;\; A = -B$$
$$1 = 2B + 2B$$
$$B = \frac{1}{4}, \qquad A = -\frac{1}{4}$$

$$\frac{1}{(u+2)(u-2)} = -\frac{1}{4(u+2)} + \frac{1}{4(u-2)}$$

The integral becomes,

$$\frac{1}{2}\int_1^2 \frac{du}{(u+2)(u-2)} = -\frac{1}{8}\int_1^2 \frac{du}{u+2} + \frac{1}{8}\int_1^2 \frac{du}{u-2}$$

$$\frac{1}{8}[-\ln(|u+2|) + \ln(|u-2|)]_1^2$$

$$= \frac{1}{8}\left[\ln\left(\left|\frac{u-2}{u+2}\right|\right)\right]_1^2$$

But $u = 2x - 3$

$$= \frac{1}{8}\left[\ln\left(\left|\frac{2x-3-2}{2x-3+2}\right|\right)\right]_1^2$$

$$= \frac{1}{8}\left[\ln\left(\left|\frac{2x-5}{2x-1}\right|\right)\right]_1^2$$

$$= \frac{1}{8}\left[\ln\left(\left|\frac{4-5}{4-1}\right|\right) - \ln\left(\left|\frac{2-5}{2-1}\right|\right)\right]$$

$$= \frac{1}{8}\left[\ln\left(\left|\frac{-1}{-3}\right|\right) - \ln\left(\left|\frac{-3}{-1}\right|\right)\right]$$

$$= \frac{1}{8}\left[\ln\left(\frac{1}{3}\right) - \ln(3)\right]$$

$$\frac{1}{8}\left[\ln\left(\frac{1}{3}\right) + \ln\left(\frac{1}{3}\right)\right]$$

$$\frac{1}{8}\ln\left(\frac{1}{9}\right)$$

The keystrokes for your *fx-991ES* Calculator are,

Confirm this result, perform the integral on your *fx* calculator,

$$\int_{1}^{2} \frac{dx}{4x^2 - 12x + 5}$$

The keystrokes are,

[∫] [1] [▭/▭] [4] [ALPHA] (X) [x^2] [−] [1] [2] [RCL] (X) [+] [5] [▼] [1] [▲] [3] [=]

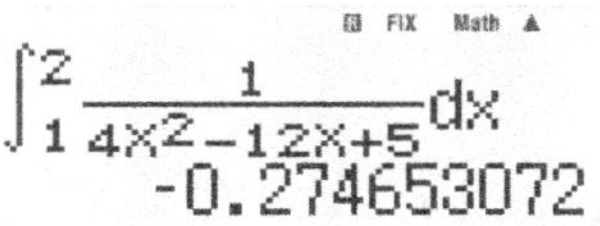

Which confirms the integration is correct. Here you will see a plot of the function,

$$f(x) = \frac{1}{4x^2 - 12x + 5}$$

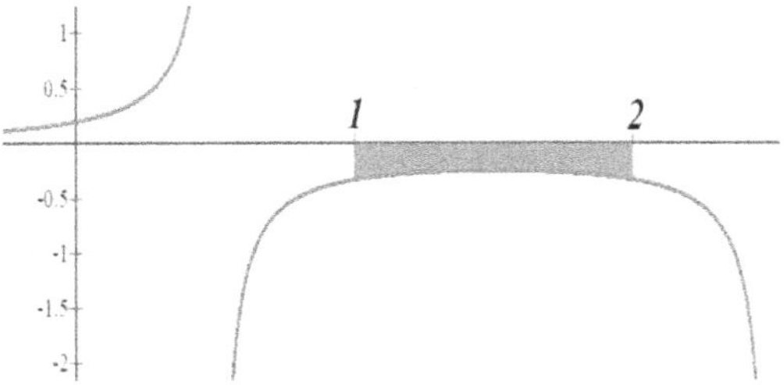

The pink area is the value of the integration. As you can see there are two asymptotes in this function, do you know what value of x they occur?

Example 5:

Perform the integration

$$\int_{1}^{2} \frac{e^{2x}}{1 + e^x} dx$$

Make the substitution,

$$u = 1 + e^x$$

Then

$$du = e^x\,dx, \qquad \text{or} \qquad dx = \frac{du}{e^x}$$

The integral becomes,

$$\int_1^2 \frac{e^{2x}}{1+e^x}\,dx = \int_1^2 \frac{e^{2x}}{u}.\frac{du}{e^x} = \int_1^2 \frac{e^x}{u}\,du$$

Since $u = 1 + e^x$, then $e^x = u - 1$

The integral becomes,

$$\int_1^2 \frac{u-1}{u}\,du$$

$$= \int_1^2 \left(1 - \frac{1}{u}\right)\,du$$

Performing the integration,

$$\int_1^2 \left(1 - \frac{1}{u}\right)\,du = [u - \ln(|u|)]_1^2$$

$$[u - \ln(|u|)]_1^2$$

Substitute back,

$$u = 1 + e^x$$

$$[1 + e^x - \ln(|1 + e^x|)]_1^2$$

This becomes,

$$[1 + e^2 - \ln(|1 + e^2|) - 1 - e^1 + \ln(|1 + e^1|)]$$

$$e^2 - e^1 + \ln\left(\frac{1 + e^1}{1 + e^2}\right)$$

The keystrokes for your *fx* calculator to perform this calculation are,

[SHIFT] ($e^{\blacksquare}$) [2] [▶] [−] [SHIFT] ($e^{\blacksquare}$) [1] [▶] [+] [ln] [$\frac{\square}{\square}$] [1] [+] [SHIFT] ($e^{\blacksquare}$) [1] [▼] [1] [+] [SHIFT] ($e^{\blacksquare}$) [2] [▶] [▶] [)] [=]

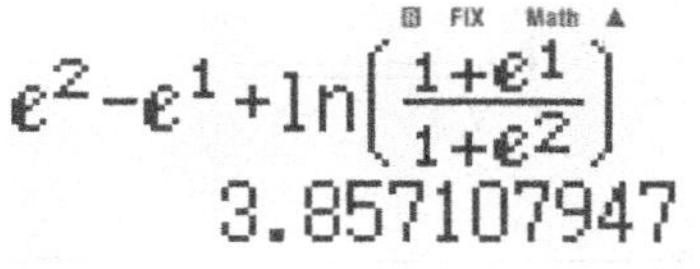

To verify this result use the integration feature on your calculator,

$$\int_1^2 \frac{e^{2x}}{1 + e^x} dx$$

The keystrokes to calculate this integral are,

[$\int_\square^\square$] [$\frac{\square}{\square}$] [SHIFT] ($e^{\blacksquare}$) [2] [ALPHA] (X) [▼] [1] [+] [SHIFT] ($e^{\blacksquare}$) [RCL] (X) [▼] [1] [▲] [2] [=]

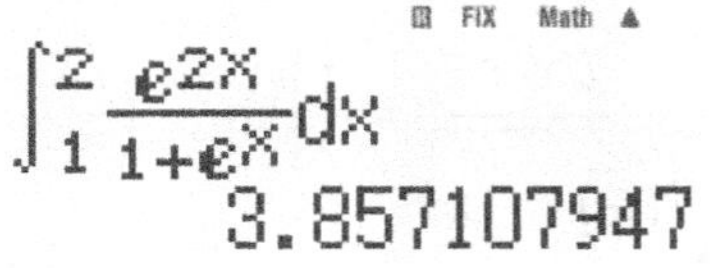

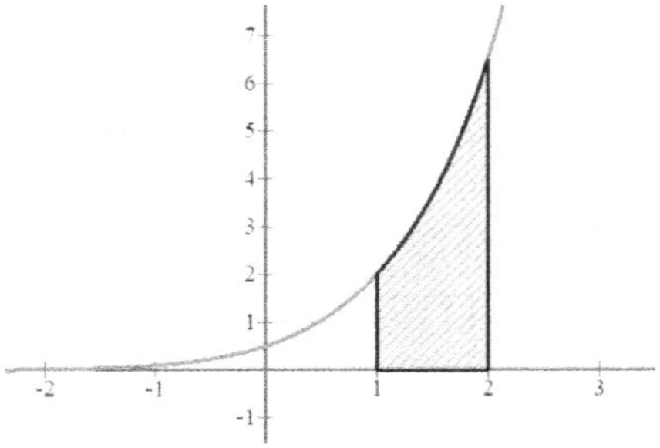

Which confirms the correctness of the integration. Here is a plot of the function.

Example 6:

Perform the integration

$$\int_{-2}^{1} \frac{dx}{\sqrt{25 - 4x^2}}$$

Express the integral as,

$$\int_{-2}^{1} \frac{dx}{\sqrt{5^2 - 4x^2}}$$

$$= \frac{1}{2} \int_{-2}^{1} \frac{dx}{\sqrt{\left(\frac{5}{2}\right)^2 - x^2}}$$

Using the well known result,

$$\int \frac{dx}{\sqrt{a^2 - x^2}} = \sin^{-1}\left(\frac{x}{a}\right), \quad a = \frac{5}{2}$$

The integral becomes,

$$= \frac{1}{2} \int_{-2}^{1} \frac{dx}{\sqrt{\left(\frac{5}{2}\right)^2 - x^2}} = \frac{1}{2}\left[\sin^{-1}\left(\frac{2x}{5}\right)\right]_{-2}^{1}$$

$$\frac{1}{2}\left[\sin^{-1}\left(\frac{2}{5}\right) - \sin^{-1}\left(-\frac{4}{5}\right)\right]$$

Perform this calculation on your *fx* calculator. First set in in radian mode ([SHIFT] [MODE] [4]). The keystrokes are,

[1] [▭/▭] [2] [▶] [(] [SHIFT] (sin^{-1}) [2] [▭/▭] [5] [▶] [)] [−] [SHIFT] (sin^{-1}) [(−)] [4] [▭/▭] [5] [▶] [)] [)] [=]

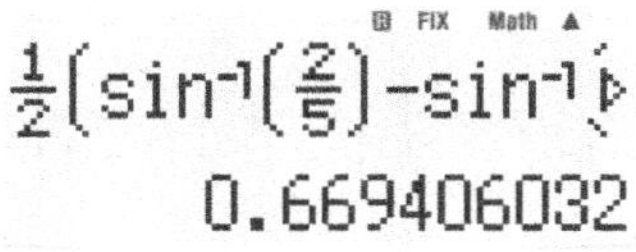

Verify the result,

$$\int_{-2}^{1} \frac{dx}{\sqrt{25-4x^2}}$$

with the following keystrokes

Which confirms the integration is correct.

Example 7:

Perform the integration

$$\int_{2}^{4} \frac{x^5}{x^3-1} dx$$

The denominator is of a higher order than the numerator, perform division.

$$\begin{array}{r|l} & x^2 \\ \hline x^3-1 & x^5 \\ & \underline{x^5 - x^2} \\ & \quad x^2 \end{array}$$

The result is,

$$x^2 + \frac{x^2}{x^3-1}$$

The integral becomes,

$$\int_2^4 \frac{x^5}{x^3-1}dx = \int_2^4 \left[x^2 + \frac{x^2}{x^3-1}\right]dx$$

$$= \int_2^4 x^2 dx + \int_2^4 \frac{x^2}{x^3-1}dx = \int_2^4 x^2 dx + \frac{1}{3}\int_2^4 \frac{3x^2}{x^3-1}dx$$

$$= \left[\frac{x^3}{3}\right]_2^4 + \frac{1}{3}\int_2^4 \frac{3x^2}{x^3-1}dx$$

$$= \left[\frac{x^3}{3}\right]_2^4 + \frac{1}{3}[\ln(|x^3-1|)]_2^4$$

$$= \frac{1}{3}[x^3 + \ln(|x^3-1|)]_2^4$$

$$= \frac{1}{3}[4^3 - 2^3 + \ln(|4^3-1|) - \ln(|2^3-1|)]$$

$$= \frac{1}{3}\left[4^3 - 2^3 + \ln\left(\left|\frac{4^3-1}{2^3-1}\right|\right)\right]$$

$$= \frac{1}{3}\left[56 + \ln\left(\left|\frac{63}{7}\right|\right)\right]$$

The keystrokes for your *fx-991ES* calculator are,

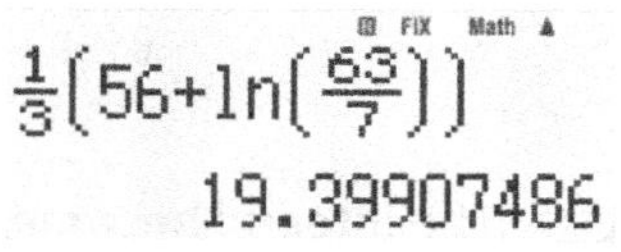

$$\int_{2}^{4} \frac{x^5}{x^3 - 1}\,dx$$

Verify the result by performing the Integration on your calculator. The keystrokes are,

[AC] [∫] [▭/▭] [ALPHA] (X) [$x^{■}$] [5] [▼] [RCL] (X) [SHIFT] (x^3) [−] [1] [▼] [2] [▲] [4] [=]

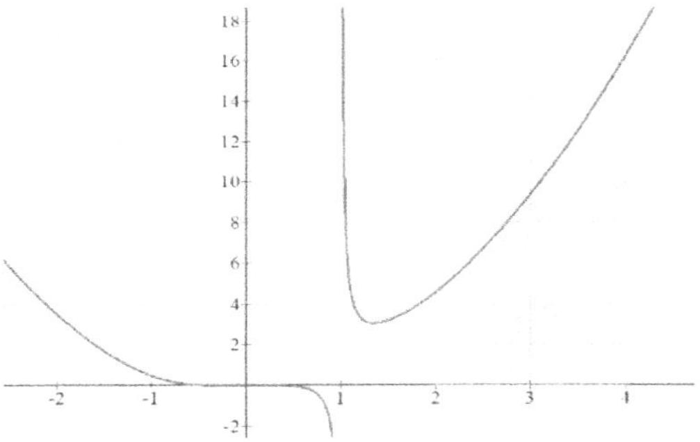

Which confirms the correct integration. A plot of the function is shown on the left together with the area of integration.

Example 8:

Calculating the dark green area enclosed by a function.

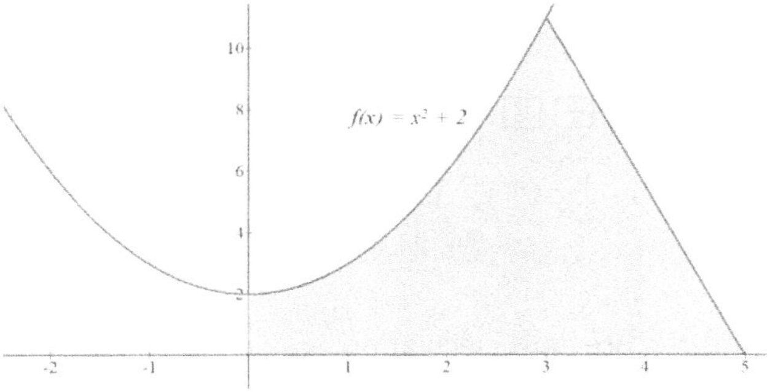

Partition the area in two sections,

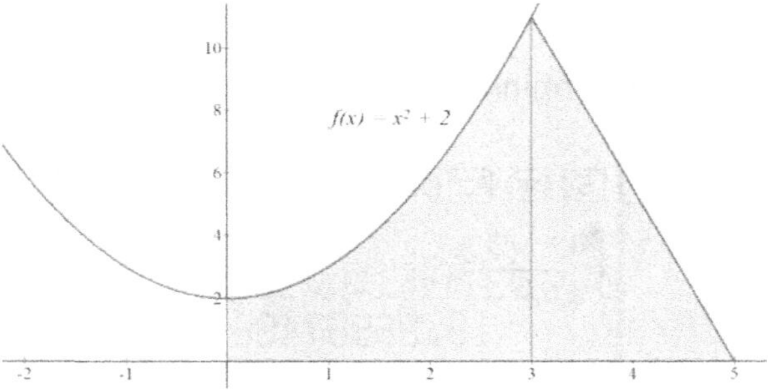

$$area = \int_0^3 (x^2 + 2)dx + \frac{1}{2} base \times height$$

Perform the integration and from the graph,

$$\left[\frac{x^3}{3} + 2x\right]_0^3 + \frac{1}{2} \times 2 \times 11$$

$$\frac{3^3}{3} + 2 \times 3 + 11 = 9 + 6 + 11 = 26$$

$$\int_0^3 (x^2 + 2)dx + \frac{1}{2} \times 2 \times 11$$

Confirm this on your fx calculator with the following keystrokes.

Which confirms the previous calculation.

Example 9:

Calculating the area enclosed by a functions.

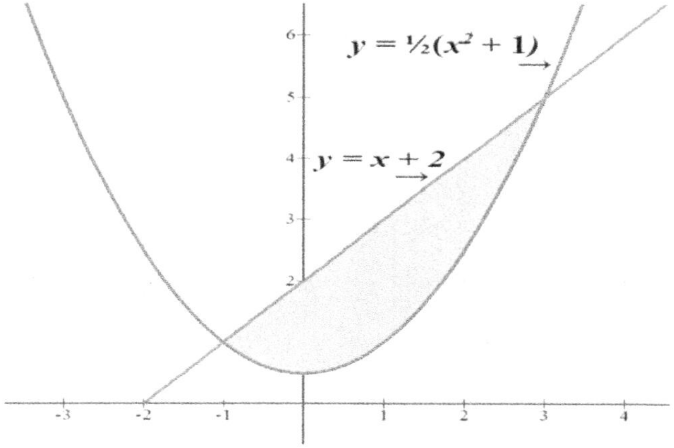

Define a triangular region,

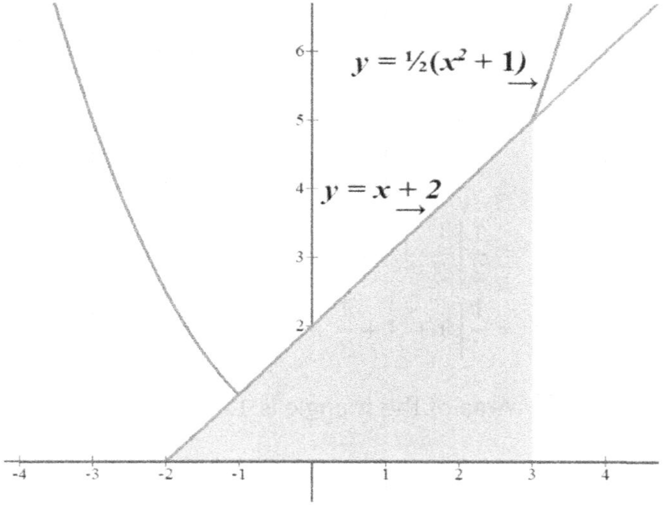

Area of triangle = $\frac{1}{2} \times base \times height$

$$=\frac{1}{2}\times 5\times 5 = 12\frac{1}{2}$$

Calculate the area under the curve,

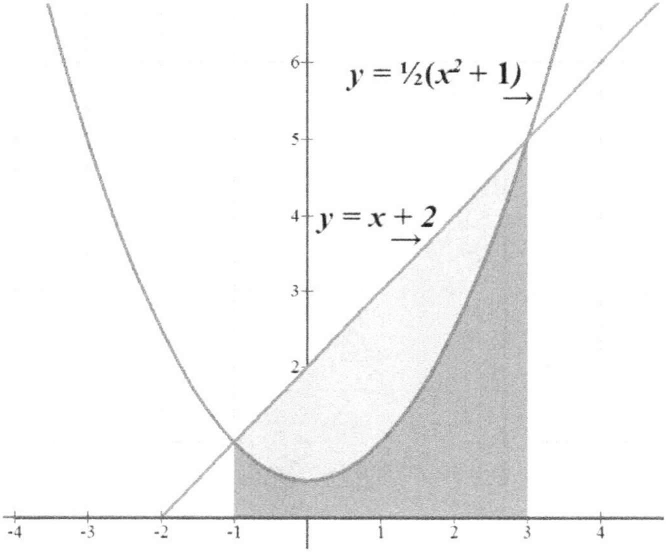

Area of grey = $\frac{1}{2}\int_{-1}^{3}(x^2+1)dx$

$$=\frac{1}{2}\left[\frac{x^3}{3}+x\right]_{-1}^{3}$$

$$=\frac{1}{2}\left[\frac{3^3}{3}+3-\frac{(-1)^3}{3}+1\right]$$

$$=\frac{1}{2}\left[9+3+\frac{1}{3}+1\right]=6\frac{2}{3}$$

Area of the triangle is $12\frac{1}{2}$

The difference is $12\frac{1}{2}-6\frac{2}{3}$

You calculator keystrokes are,

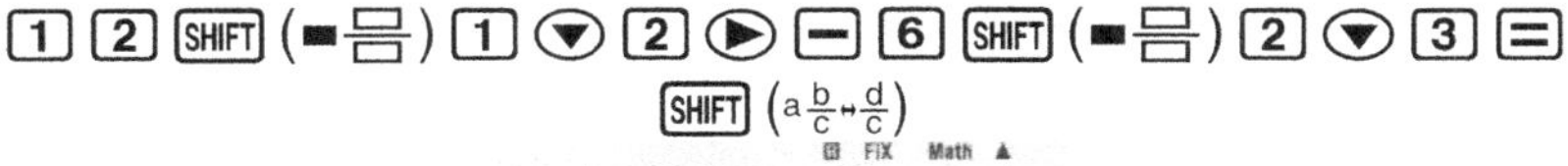

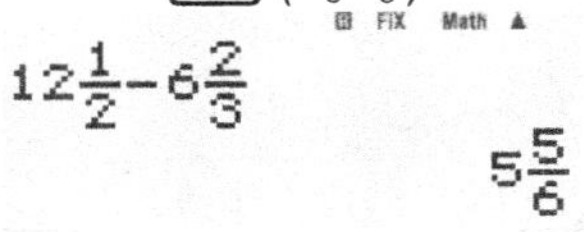

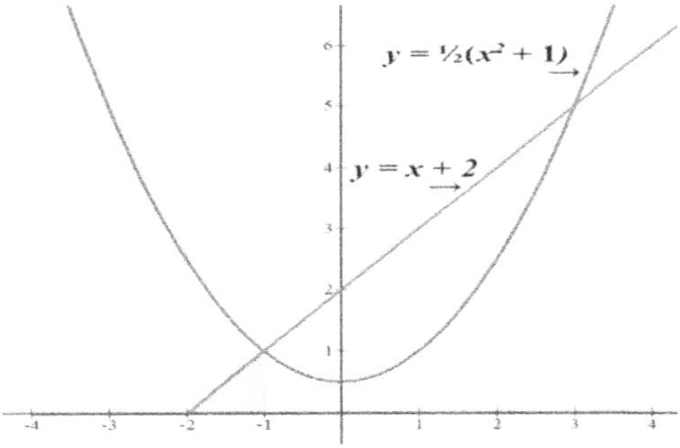

The area of the small triangle in light blue is = $\frac{1}{2}$. Need to subtract $\frac{1}{2}$ from $5\frac{5}{6}$. Additional keystrokes,

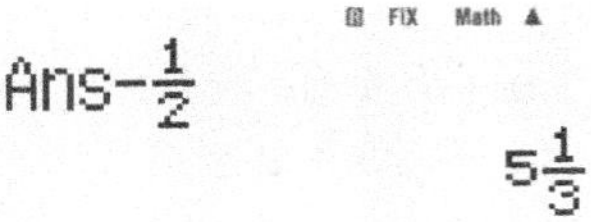

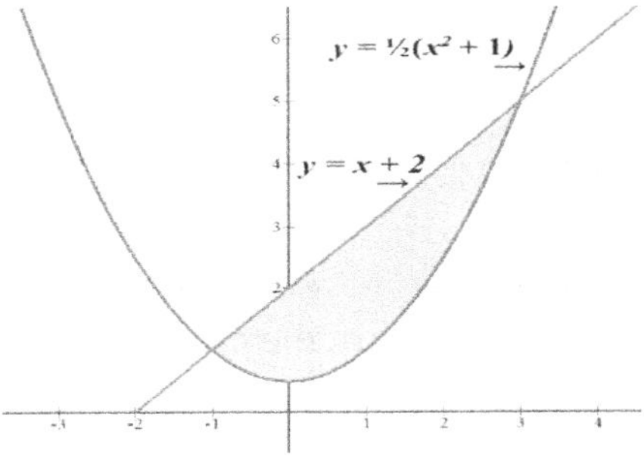

The area enclosed by the functions is therefore 5.333 square units.

Example 10:

Perform the following integration,

$$\int_3^4 \frac{2x+4}{(x-2)(x^2+4)}\,dx$$

Whenever you see an integral of this form you need to use partial fractions.

$$\frac{2x+4}{(x-2)(x^2+4)} = \frac{a}{(x-2)} + \frac{bx+c}{(x^2+4)}$$

This becomes,

$$2x+4 = a(x^2+4) + (bx+c)(x-2)$$

$$= ax^2 + 4a + bx^2 - 2bx + cx - 2c$$

Equate coefficients

$$0 = a + b$$
$$2 = -2b + c$$
$$4 = 4a - 2c$$

Write them in this form

$$a + b + 0c = 0$$
$$0a - 2b + c = 2$$
$$2a + 0b - c = 2$$

Express as a matrix

$$\begin{pmatrix} 1 & 1 & 0 \\ 0 & -2 & 1 \\ 2 & 0 & -1 \end{pmatrix}\begin{pmatrix} a \\ b \\ c \end{pmatrix} = \begin{pmatrix} 0 \\ 2 \\ 2 \end{pmatrix}$$

$$\boldsymbol{AX} = \boldsymbol{B}$$

To find the values of the coefficients, need to perform the calculation,

$$\boldsymbol{X} = \boldsymbol{A}^{-1}\boldsymbol{B}$$

$$\begin{pmatrix} a \\ b \\ c \end{pmatrix} = \begin{pmatrix} 1 & 1 & 0 \\ 0 & -2 & 1 \\ 2 & 0 & -1 \end{pmatrix}^{-1} \begin{pmatrix} 0 \\ 2 \\ 2 \end{pmatrix}$$

The keystrokes for your *fx* calculator, first select the matrix mode ([MODE] [6]). Select [1] [1]. Enter the values from the matrix,

[1] [=] [1] [=] [0] [=] [0] [=] [(−)] [2] [=] [1] [=] [2] [=] [0] [=] [(−)] [1] [=] [AC]

Enter the following keystrokes,

[SHIFT] (MATRIX) [2] [2] [3]

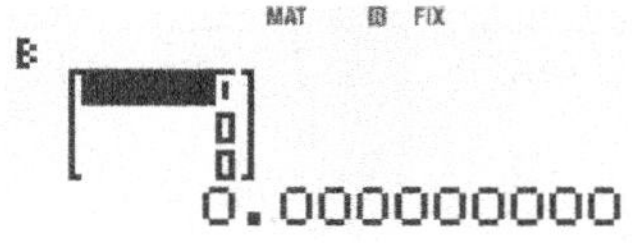

Enter the data into the column matrix

[0] [=] [2] [=] [2] [=] [AC]

Need to perform the calculation,

$$\boldsymbol{X} = \boldsymbol{A}^{-1}\boldsymbol{B}$$

Enter the keystrokes,

[SHIFT] (MATRIX) [3] [x^{-1}] [×] [SHIFT] (MATRIX) [4]

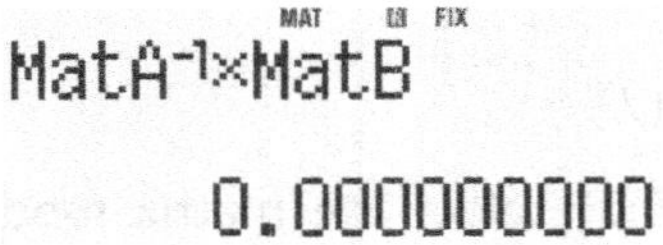

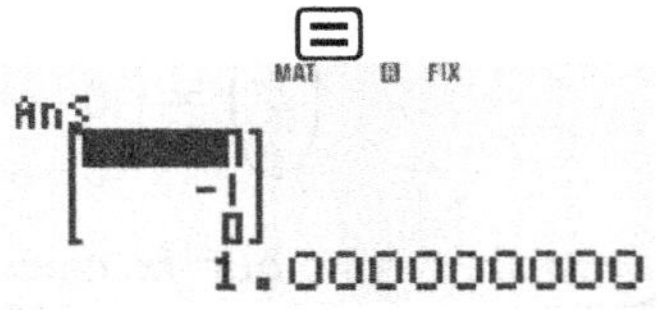

$$a = 1, b = -1 \text{ and } c = 0$$

The partial fraction becomes,

$$\frac{2x+4}{(x-2)(x^2+4)} = \frac{1}{(x-2)} + \frac{-x}{(x^2+4)}$$

The integral becomes,

$$\int_3^3 \frac{2x+4}{(x-2)(x^2+4)} dx$$

$$= \int_3^4 \frac{1}{(x-2)} dx - \int_3^4 \frac{x}{(x^2+4)} dx$$

$$= \int_3^4 \frac{1}{(x-2)} dx - \frac{1}{2}\int_3^4 \frac{2x}{(x^2+4)} dx$$

$$= [\ln(|x-2|)]_3^4 - \frac{1}{2}[\ln(|x^2+4|)]_3^4$$

$$= [\ln(|4-2|) - \ln(|3-2|)] - \frac{1}{2}[\ln(|4^2+4|) - \ln(|3^2+4|)]$$

$$= [\ln(|2|) - \ln(|1|)] - \frac{1}{2}[\ln(|20|) - \ln(|13|)]$$

The keystrokes for this calculation (calculator in COMP mode [MODE] [1])

[ln] [2] [)] [−] [1] [▭/▭] [2] [▶] [(] [ln] [2] [0] [)] [−]
[ln] [1] [3] [)] [)] [=]

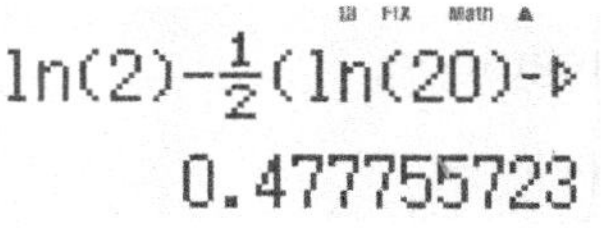

$$\int_3^4 \frac{2x+4}{(x-2)(x^2+4)}dx$$

Confirm the result with the following keystrokes on your calculator,

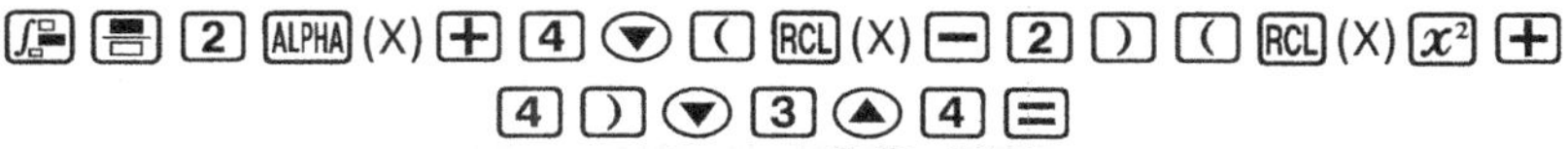

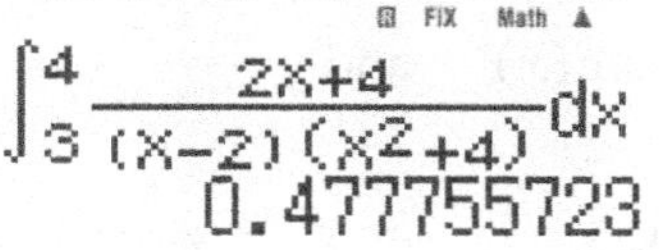

Which confirms the integration is correct. A plot of the function and in the integrated area is shown here,

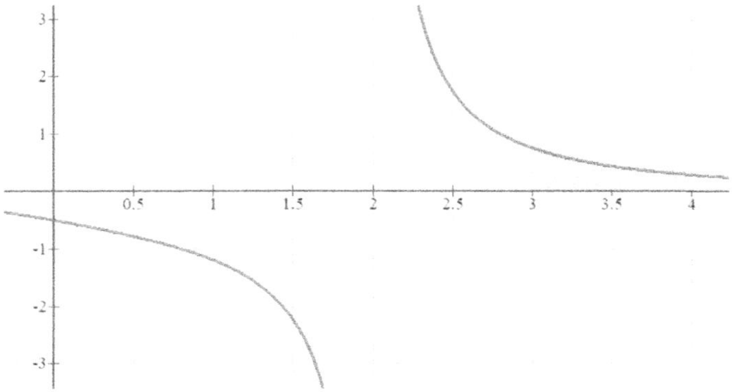

You will observe an asymptote at $x = 2$.

Example 11:

Perform the following integration,

$$\int_{\pi/6}^{\pi/3} \sec^3(x)\tan(x)\,dx$$

Express as,

$$\int_{\pi/6}^{\pi/3} \frac{1}{\cos^3(x)}\frac{\sin(x)}{\cos(x)}\,dx = \int_{\pi/6}^{\pi/3} \frac{\sin(x)}{\cos^4(x)}\,dx$$

Let

$$u = \cos(x)$$

Then

$$\frac{du}{dx} = -\sin(x) \quad \text{and} \quad -\frac{du}{\sin(x)} = dx$$

The integral becomes

$$\int_{\pi/6}^{\pi/3} \frac{\sin(x)}{\cos^4(x)}\,dx = -\int_{\pi/6}^{\pi/3} \frac{\sin(x)}{\cos^4(x)}\frac{du}{\sin(x)}$$

This becomes,

$$-\int_{\frac{\pi}{6}}^{\frac{\pi}{3}} \frac{du}{u^4} = -\int_{\frac{\pi}{6}}^{\frac{\pi}{3}} u^{-4}\,du = \left[\frac{u^{-3}}{3}\right]_{\pi/6}^{\pi/3}$$

$$= \frac{1}{3}\left[\frac{1}{\cos^3(x)}\right]_{\pi/6}^{\pi/3} = \frac{1}{3}\left[\frac{1}{\cos^3\left(\frac{\pi}{3}\right)} - \frac{1}{\cos^3\left(\frac{\pi}{6}\right)}\right]$$

The keystrokes for your *fx* calculator to evaluate this are

[SHIFT] (SETUP) [4] [1] [fraction] [3] [▶] [(] [1] [fraction] [cos] [SHIFT] (π) [fraction] [3] [▶] [)] [SHIFT] (x^3) [▶] [−] [1] [fraction] [cos] [SHIFT] (π) [fraction] [6] [▶] [)] [SHIFT] (x^3) [▶] [)] [=]

[S⇔D]

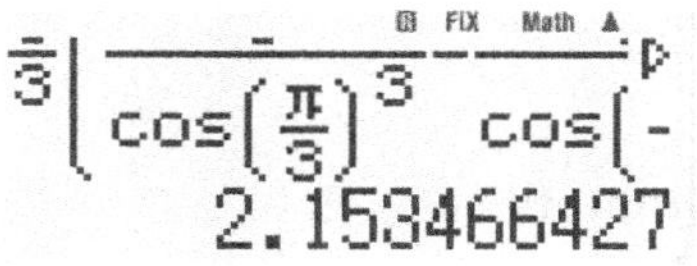

Confirm this result,

$$\int_{\pi/6}^{\pi/3} \sec^3(x)\tan(x)\,dx = \int_{\pi/6}^{\pi/3} \frac{\tan(x)}{\cos^3(x)}\,dx$$

The keystrokes for performing this Integral on your fx calculator,

[∫] [tan] [ALPHA] (X) [ )] [▭/▭] [cos] [RCL] (X) [ )] [SHIFT] (x^3) [▼] [SHIFT] (π) [÷] [6] [▲]
[SHIFT] (π) [÷] [3] [=]

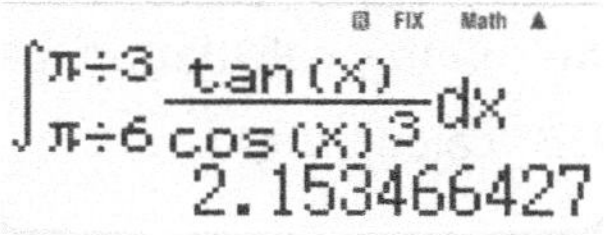

Example 12:

Perform the integration,

$$\int_0^{9\pi/2} x\sin(x)\,dx$$

Make sure your calculator is in radian mode. When integrating by parts use the following,

$$\int u\frac{dv}{dx}dx = uv - \int v\frac{du}{dx}dx$$

Let $u = x$ and $\frac{dv}{dx} = \sin(x)$, then

$$\frac{du}{dx} = 1 \text{ and } v = \int \sin(x)\,dx = -\cos(x)$$

$$\int_0^{9\pi/2} x\sin(x)\,dx = -[x\cos(x)]_0^{9\pi/2} + \int_0^{9\pi/2} \cos(x)dx$$

Complete the integration,

$$\int_0^{9\pi/2} x\sin(x)\,dx = -[x\cos(x)]_0^{9\pi/2} + [\sin(x)]_0^{9\pi/2}$$

This becomes,

$$-\frac{9\pi}{2}\cos\left(\frac{9\pi}{2}\right) + \sin\left(\frac{9\pi}{2}\right)$$

Since $\frac{9\pi}{2}$ appears three times in this calculation, use the following keystrokes for your *fx-991ES* calculator,

[9] [SHIFT] (π) [▭/▭] [2] [▶] [ALPHA] (:) [(−)] [Ans] [×] [cos] [Ans] [)] [+] [sin] [Ans] [)] [=]

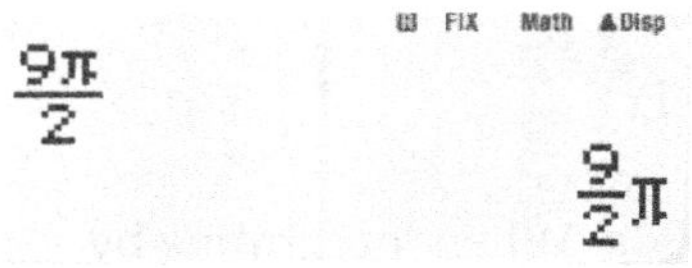

[=]

$$\int_0^{9\pi/2} x\sin(x)\,dx$$

To verify the result, perform the calculation directly on your fx calculator. Use the keystrokes,

[∫▭] [ALPHA] (X) [sin] [RCL] (X) [)] [▼] [0] [▲] [9] [SHIFT] (π) [÷] [2] [=]

You are probably surprised to see the answer of 1. On the next page there is a plot of the function and the area of integration.

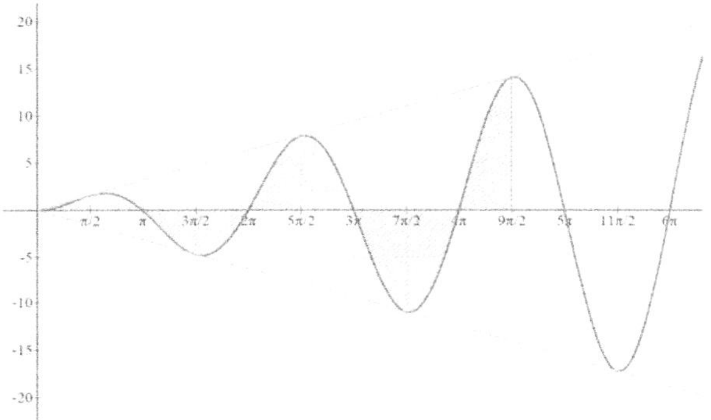

The value of 9π/2 was chosen to give exactly an area of value 1.

Example 13:

Evaluate the integral

$$\int_{0}^{3} x\ln(1+x)\,dx$$

Perform integration by parts

$$\int u\frac{dv}{dx}dx = uv - \int v\frac{du}{dx}\,dx$$

Let $u = \ln(1+x)$ and $\frac{dv}{dx} = x$

Then $\frac{du}{dx} = \frac{1}{x+1}$ and $v = \int x\,dx\ = \frac{x^2}{2}$

$$\int_{0}^{3} x\ln(1+x)\,dx = \left[\frac{x^2}{2}\ln(1+x)\right]_0^3 - \frac{1}{2}\int_{0}^{3}\frac{x^2}{x+1}dx$$

In the integral need to divide the numerator by the denominator.

$$\begin{array}{r}
x \;-1 \\
x+1\,|\,x^2 \qquad \\
\underline{x^2+x} \\
-x \\
\underline{-x-1} \\
1
\end{array}$$

The result is,

$$\frac{x^2}{x+1} = x - 1 + \frac{1}{x+1}$$

The integral becomes

$$\frac{1}{2}\int_0^3 \frac{x^2}{x+1}\,dx = \frac{1}{2}\int_0^3 \left[x - 1 + \frac{1}{x+1}\right] dx$$

Performing the integration,

$$= \frac{1}{2}\left[\frac{x^2}{2} - x + \ln(|x+1|)\right]_0^3$$

The complete integral is,

$$\int_0^3 x\ln(1+x)\,dx = \left[\frac{x^2}{2}\ln(1+x)\right]_0^3 - \frac{1}{2}\left[\frac{x^2}{2} - x + \ln(|x+1|)\right]_0^3$$

$$= \left[\frac{x^2}{2}\ln(1+x) - \frac{x^2}{4} + \frac{x}{2} - \frac{1}{2}\ln(|x+1|)\right]_0^3$$

$$= \frac{1}{2}\left[(x^2-1)\ln(|x+1|) - \frac{x^2}{2} + x\right]_0^3$$

Leaving,

$$\int_0^3 x\ln(x+1)dx = \frac{1}{2}\left[(3^2-1)\ln(4) - \frac{3^2}{2} + 3\right]$$

The keystrokes for your *fx-991ES* calculator are,

[1] [▭/▭] [2] [▶] [(] [8] [ln] [4] [)] [−] [4] [·] [5] [+] [3] [)] [=]

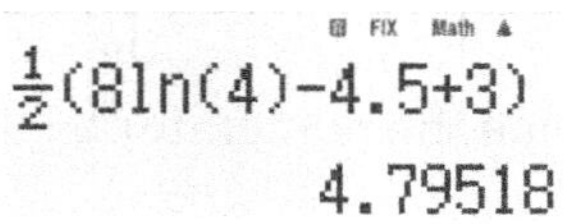

$$\int_{0}^{3} x \ln(x + 1)dx$$

Verify this using the integration feature on your calculator with the following keystrokes,

[∫□] [ALPHA] (X) [ln] [(] [RCL] (X) [+] [1] [)] [▼] [0] [▲] [3] [=]

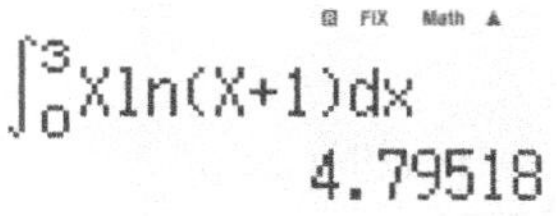

Which confirms the integration is correct. A plot of the function and the integrated area is shown below.

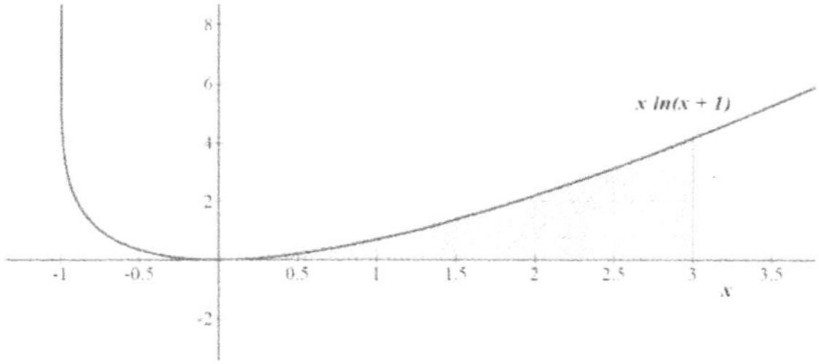

Example 14:

Evaluate the integral,

$$\int_{-3}^{0} \frac{x}{\sqrt{1 - 4x}} dx$$

Make the substitution

$$u = 1 - 4x$$
$$du = -4dx$$
$$dx = -\frac{du}{4}$$

Also,

$$x = \frac{1-u}{4}$$

The integral becomes,

$$\int_{-3}^{0} \frac{x}{\sqrt{1-4x}}\, dx = \int_{-3}^{0} \left(\frac{1-u}{4}\right) \frac{1}{\sqrt{u}} \left(-\frac{du}{4}\right)$$

This becomes,

$$-\frac{1}{16} \int_{-3}^{0} \frac{1-u}{\sqrt{u}}\, du$$

Expanding

$$-\frac{1}{16} \int_{-3}^{0} \left[\frac{1}{\sqrt{u}} - \frac{u}{\sqrt{u}}\right] du$$

This integral becomes

$$-\frac{1}{16} \int_{-3}^{0} \left[\frac{1}{\sqrt{u}} - \frac{u}{\sqrt{u}}\right] du = -\frac{1}{16} \int_{-3}^{0} \left[u^{-\frac{1}{2}} - u^{\frac{1}{2}}\right] du$$

Perform the integration,

$$= -\frac{1}{16} \int_{-3}^{0} \left[u^{-\frac{1}{2}} - u^{\frac{1}{2}}\right] du = -\frac{1}{16} \left[\frac{u^{\frac{1}{2}}}{1/2} - \frac{u^{\frac{3}{2}}}{3/2}\right]_{-3}^{0}$$

This becomes,

$$-\frac{1}{8} \left[\sqrt{u} - \frac{\sqrt{u^3}}{3}\right]_{-3}^{0}$$

Now $u = 1 - 4x$

$$-\frac{1}{8} \left[\sqrt{1-4x} - \frac{\sqrt{(1-4x)^3}}{3}\right]_{-3}^{0}$$

This becomes,

$$-\frac{1}{8}\left[\sqrt{1}-\frac{\sqrt{1}}{3}-\sqrt{1-4\times(-3)}+\frac{\sqrt{(1-4(-3))^3}}{3}\right]$$

$$=-\frac{1}{8}\left[1-\frac{1}{3}-\sqrt{13}+\frac{\sqrt{13^3}}{3}\right]$$

$$=-\frac{1}{8}\left[1-\frac{1}{3}-\sqrt{13}+\frac{\sqrt{13^3}}{3}\right]$$

The keystrokes for your calculator are,

(−) 1 [fraction] 8 ▶ (2 [fraction] 3 ▶ − √ 1 3 ▶ + √ 1 3 SHIFT (x^3) ▶ [fraction] 3 ▶) = S⇔D

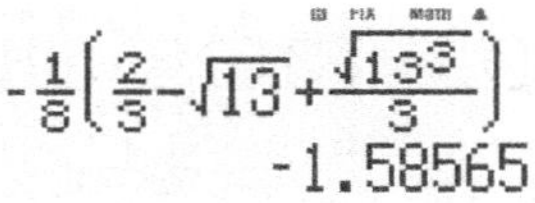

$$\int_{-3}^{0}\frac{x}{\sqrt{1-4x}}\,dx$$

Verify the result with the following Keystrokes on your calculator,

[∫] ALPHA (X) [fraction] √ 1 − 4 RCL (X) ▼ (−) 3 ▲ 0 =

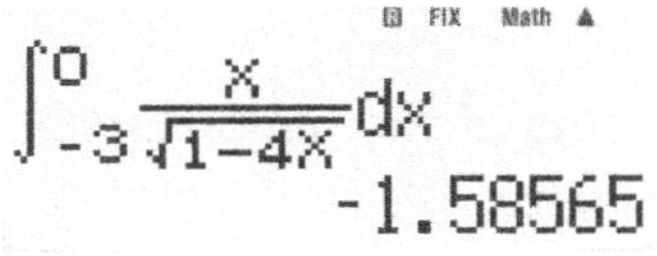

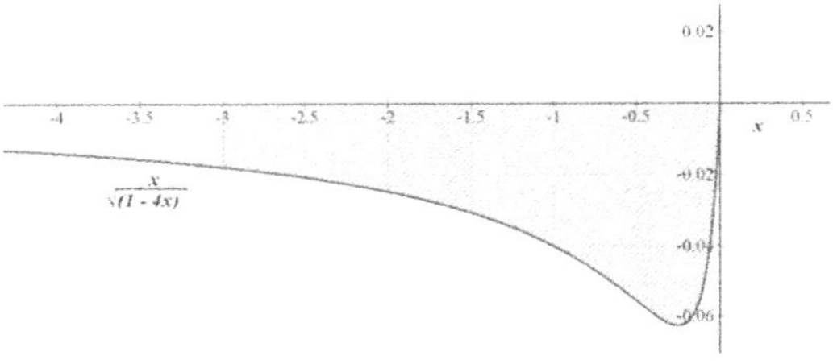

Which confirms the integration is correct. A plot of the function and integrated area is shown on the left.

Example 15:

Evaluate the integral,

$$\int_{-0.5}^{1} \frac{dx}{x^3 + 6x^2 + 11x + 6}$$

Factorise the denominator. Your *fx-991ES* can find the roots of the denominator. Use the Equation mode (MODE 5 4).

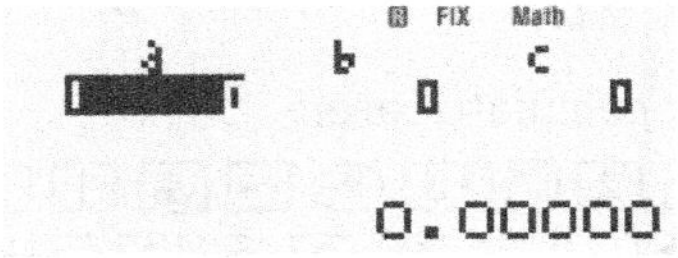

Enter the coefficient values,

1 = 6 = 1 1 = 6 =

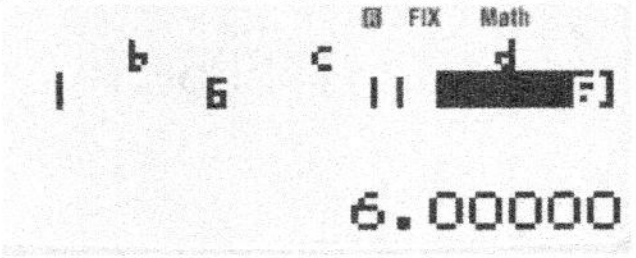

=

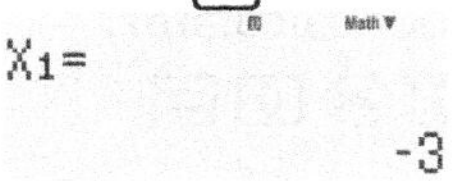

=

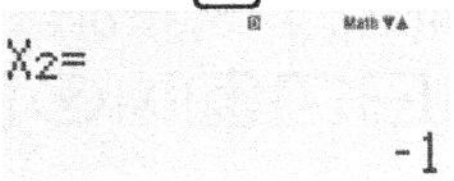

=

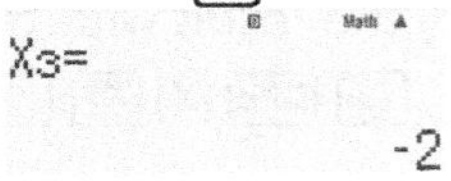

You can see the three roots of the denominator. Write the integral with a factorised denominator,

$$\int_{-0.5}^{1} \frac{dx}{x^3 + 6x^2 + 11x + 6} = \int_{-0.5}^{1} \frac{dx}{(x+3)(x+2)(x+1)}$$

Express the denominator as partial fractions,

$$\frac{1}{(x+3)(x+2)(x+1)} = \frac{a}{(x+3)} + \frac{b}{(x+2)} + \frac{c}{(x+1)}$$

$$1 = a(x+2)(x+1) + b(x+3)(x+1) + c(x+3)(x+2)$$

Expand the contents of the brackets,

$$\begin{aligned} 1 = a(x^2 + x + 2x + 2) \\ + b(x^2 + x + 3x + 3) \\ + c(x^2 + 2x + 3x + 6) \end{aligned}$$

$$1 = a(x^2 + 3x + 2) + b(x^2 + 4x + 3) + c(x^2 + 5x + 6)$$

Equate coefficients,

$$\begin{aligned} 0 &= a + b + c \\ 0 &= 3a + 4b + 5c \\ 1 &= 2a + 3b + 6c \end{aligned}$$

Express in terms of matrices

$$\begin{pmatrix} 1 & 1 & 1 \\ 3 & 4 & 5 \\ 2 & 3 & 6 \end{pmatrix} \begin{pmatrix} a \\ b \\ c \end{pmatrix} = \begin{pmatrix} 0 \\ 0 \\ 1 \end{pmatrix}$$

We require the coefficients,

$$\begin{pmatrix} a \\ b \\ c \end{pmatrix} = \begin{pmatrix} 1 & 1 & 1 \\ 3 & 4 & 5 \\ 2 & 3 & 6 \end{pmatrix}^{-1} \begin{pmatrix} 0 \\ 0 \\ 1 \end{pmatrix}$$

$$\begin{pmatrix} a \\ b \\ c \end{pmatrix} = \boldsymbol{B}^{-1}\boldsymbol{C}$$

Use the matrix features on your *fx* calculator to perform this calculation (MODE 6 2 1).

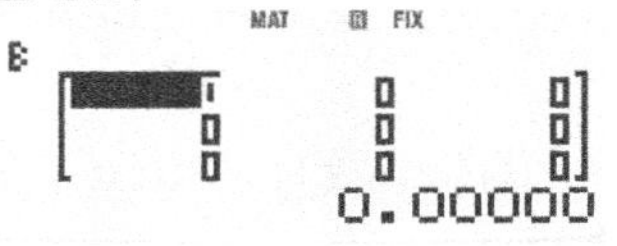

Enter the keystrokes for matrix ***B***,

[1] [=] [1] [=] [1] [=] [3] [=] [4] [=] [5] [=] [2] [=] [3] [=] [6] [=]

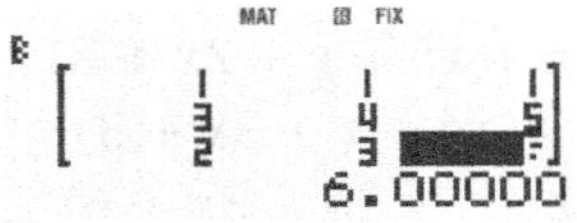

Enter the keystrokes for matrix ***C***,

[AC] [SHIFT] (MATRIX) [2] [3] [3]   [0] [=] [0] [=] [1] [=]

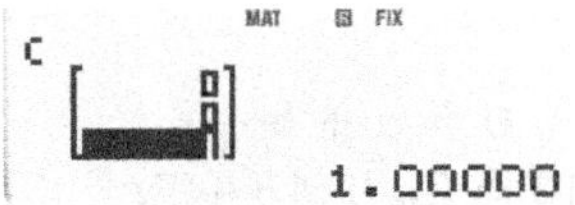

[AC] [SHIFT] (MATRIX) [4] [x^{-1}] [×] [SHIFT] (MATRIX) [5]

[=]

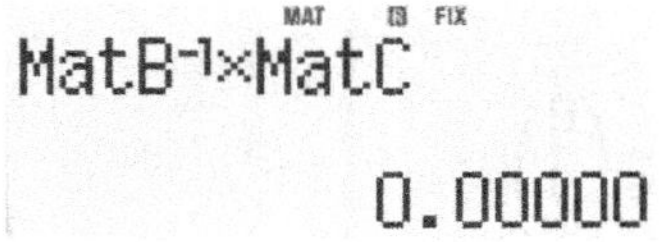

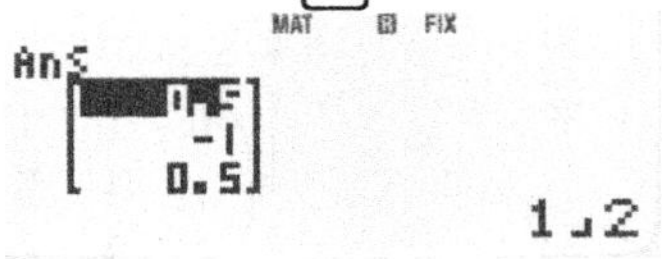

You can see the values of the coefficients in the Ans matrix, therefore,

$$\frac{1}{(x+3)(x+2)(x+1)} = \frac{1}{2(x+3)} - \frac{1}{(x+2)} + \frac{1}{2(x+1)}$$

The integral becomes,

$$\frac{1}{2}\int_{-0.5}^{1} \frac{dx}{x+3} - \int_{-0.5}^{1} \frac{dx}{x+2} + \frac{1}{2}\int_{-0.5}^{1} \frac{dx}{x+1}$$

$$= \left[\frac{1}{2}\ln(x+3) - \ln(x+2) + \frac{1}{2}\ln(x+1)\right]_{-0.5}^{1}$$

$$= \left[\frac{1}{2}\ln\{(x+3)(x+1)\} - \ln(x+2)\right]_{-0.5}^{1}$$

$$= \left[\frac{1}{2}\ln\{(1+3)(1+1)\} - \ln(1+2)\right]$$

$$- \left[\frac{1}{2}\ln\{(-0.5+3)(-0.5+1)\} - \ln(-0.5+2)\right]$$

$$= \left[\tfrac{1}{2}\ln\{(4)(2)\} - \ln(3)\right] - \left[\tfrac{1}{2}\ln\{(2.5)(0.5)\} - \ln(1.5)\right]$$

$$= \tfrac{1}{2}\ln\{8\} - \ln(3) - \tfrac{1}{2}\ln\{1.25\} + \ln(1.5)$$

$$= \frac{1}{2}\ln\left(\frac{8}{1.25}\right) + \ln\left(\frac{1.5}{3}\right)$$

Perform this calculation on your *fx* calculator. Enter the following keystrokes, (MODE 1)

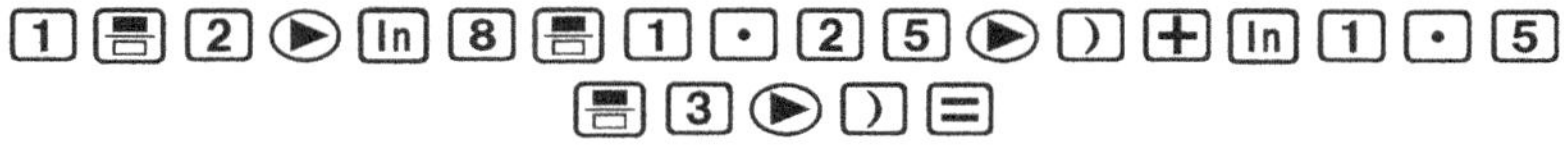

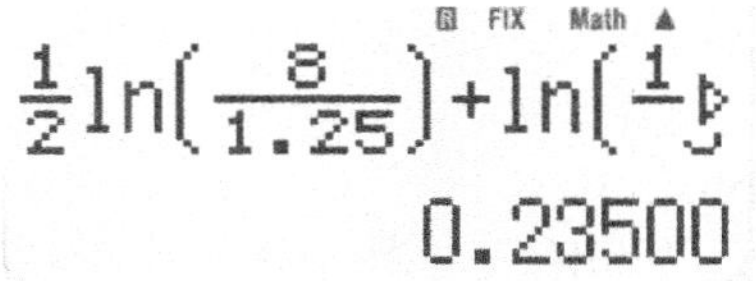

The original equation,

$$\int_{-0.5}^{1} \frac{dx}{x^3 + 6x^2 + 11x + 6}$$

To verify the correctness of the integration, enter the keystrokes for this integration,

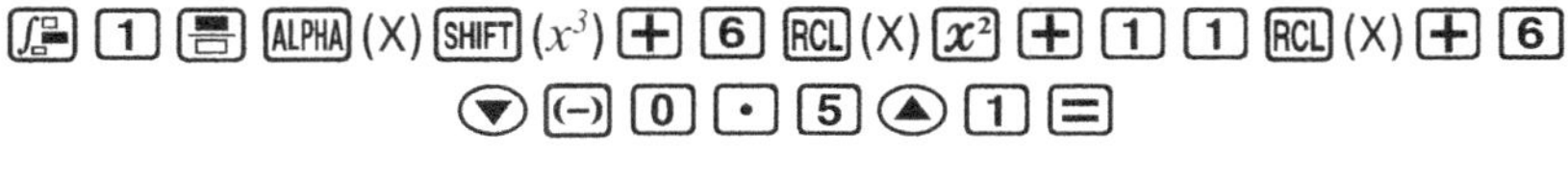

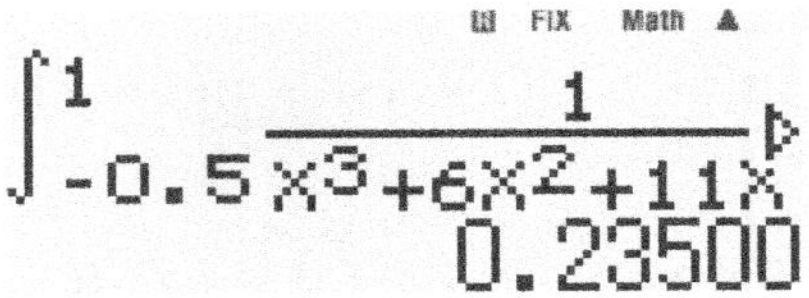

This confirms the integration is correct. A plot of the cubic function is shown below.

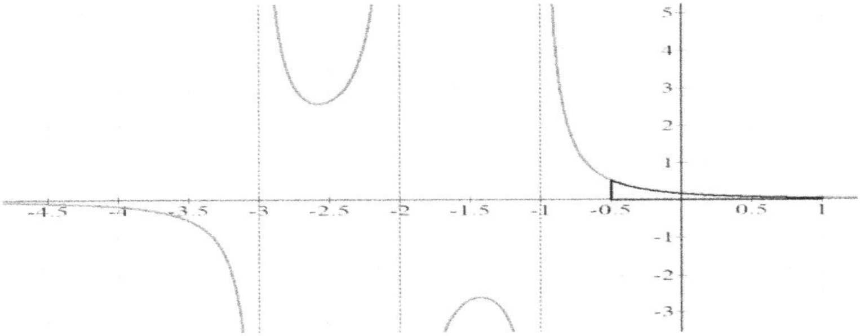

You will note there are three asymptotes, also shown is the area of integration.

Example 15:

Perform the integration,

$$\int \frac{[3\sin(x) - 2]\cos(x)}{5 - \cos^2(x) - 4\sin(x)} dx$$

Make the substitution,

$$u = \sin(x), \quad dt = \cos(x)dx$$

$$\int \frac{[3u - 2]\cos(x)}{5 - (1 - u^2) - 4u} \frac{du}{\cos(x)}$$

This becomes,

$$\int \frac{3u - 2}{u^2 - 4u + 4} du$$

The denominator is factorised to become,

$$\int \frac{3u-2}{(u-2)^2}\,du$$

We can use partial fractions,

$$\frac{3u-2}{(u-2)^2} = \frac{A}{(u-2)} + \frac{B}{(u-2)^2}$$

Multiply through by $(u-2)^2$ to give,

$$3u-2 = A(u-2) + B = uA - 2A + B$$

Equate coefficients,

$$3 = A$$
$$-2 = -2A + B$$

which leaves *A* = 3 and *B* = 4. The integral becomes,

$$3\int \frac{du}{u-2} + 4\int \frac{du}{(u-2)^2}$$

In the second integral, let $v = u - 2$, the integral becomes,

$$3\int \frac{du}{u-2} + 4\int \frac{dv}{v^2}$$

Performing the integrations,

$$3\ln|u-2| - \frac{4}{v} = 3\ln|u-2| - \frac{4}{u-2}$$

Replace u by $\sin(x)$, to leave,

$$\int \frac{[3\sin(x)-2]\cos(x)}{5-\cos^2(x)-4\sin(x)}\,dx = 3\ln|\sin(x)-2| - \frac{4}{\sin(x)-2} + c$$

Determine the value of the integral,

$$\int_{2}^{12} \frac{[3\sin(x) - 2]\cos(x)}{5 - \cos^2(x) - 4\sin(x)} dx$$

Using the derived results, this becomes,

$$\left[3\ln|\sin(x) - 2| - \frac{4}{\sin(x) - 2}\right]_2^{12}$$

We shall use the M in your *fx-991ES* calculator. First empty M with the keystrokes [0] [=] [SHIFT] (STO)(M), put the calculator in radian mode [SHIFT] (SETUP) [4], then enter the keystrokes,

[AC] [3] [ln] [SHIFT] (Abs) [sin] [ALPHA] (X) [)] [−] [2] [▶] [)] [−] [4] [fraction] [sin] [RCL] (X) [)] [−] [2]

[CALC]
X? [1] [2] [=]

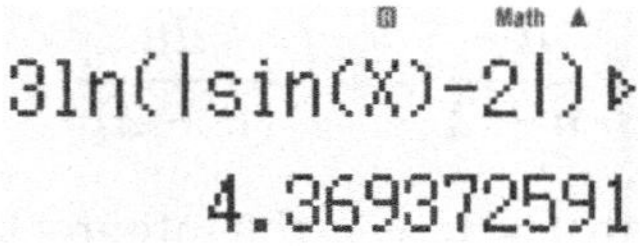

[M+] [▲] [CALC]
X? [2] [=]

[SHIFT] (M-) [RCL] (M)

To confirm the result, enter the following keystrokes into your calculator for the integral,

$$\int_{2}^{12} \frac{[3\sin(x) - 2]\cos(x)}{5 - \cos^2(x) - 4\sin(x)}\,dx$$

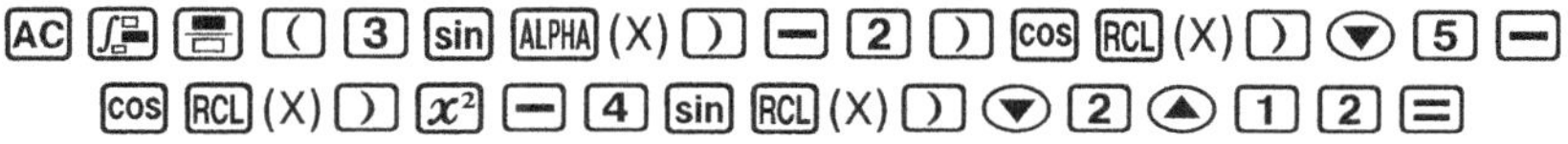

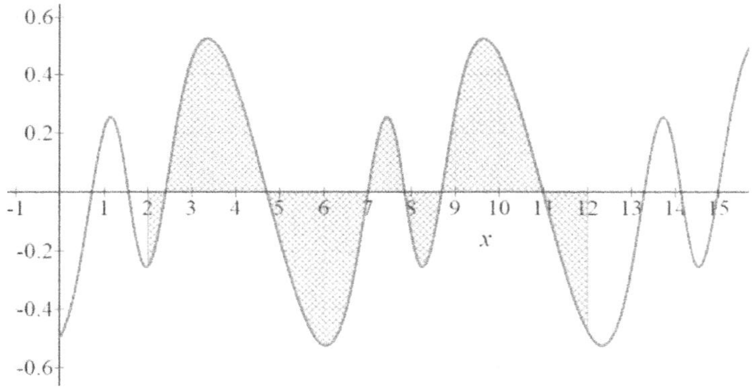

This calculation takes several seconds to perform. It confirms the correctness of the result. A graph of the function and the area of integration is shown here.

Made in the USA
Monee, IL
04 November 2026

67377527R00049